Materials Characterization

Materials Characterization

Modern Methods and Applications

edited by

Narayanaswami Ranganathan

CRC Press
Taylor & Francis Group
Boca Raton London New York

CRC Press is an imprint of the
Taylor & Francis Group, an informa business

Materials Characterization: Modern Methods and Applications

First published 2015 Pan Stanford Publishing Pte. Ltd.

Published 2022 by CRC Press
Taylor & Francis Group
6000 Broken Sound Parkway NW, Suite 300
Boca Raton, FL 33487-2742

ISBN 13: 978-981-4613-06-4 (hbk)
ISBN 13: 978-981-4613-07-1 (ebk)

Visit the Taylor & Francis Web site at
http://www.taylorandfrancis.com

and the CRC Press Web site at
http://www.crcpress.com

British Library Cataloguing-in-Publication Data
A catalogue record for this book is available from the British Library.

Contents

Preface

Modern materials are put to use in harsh environments and high temperatures, and the determination of material properties has become a key issue. The material mechanical properties are governed by the microstructure and other physical properties. Depending on the kind of application, one is interested in local mechanical properties, typically at the surface or in the layers immediately underneath, while other applications require the knowledge of the global or bulk behavior.

This book, which is a result of a coordinated effort by researchers from five different countries, addresses the methods of determining local and global mechanical properties of a variety of materials: metals, plastics, rubber, and ceramics.

Chapter 1 is a comprehensive treatment of the nanoindentation technique, treating the basic principles, contact mechanics, and various examples of surface properties determination. While conventional properties like the hardness and the reduced modulus are discussed, more delicate tests such as nanowear, nanoimpact, and micropillar compression tests are also treated in detail, with different relevant applications.

Chapter 2 treats surface property changes in selected polymers and rubbers, measured by nanoindentation. Different aspects such as surface gradient of vulcanizate crosslinking, surface migration of low molecular weight components of rubber, and surface segregation in polymer blends are discussed.

Chapter 3 treats the static and wear resistance of dental composites with emphasis on new composites with enhanced behavior like improving the mechanical properties of fillings, resulting in longer lifetime, and by providing bactericidal function to composites.

Chapter 4 treats the global and local properties of a lead free solder. It is shown that the local properties, such as hardness, modulus, and creep are affected by the presence of a specific microstructure in the solder ball and the existence of local precipitates. The local properties can be quite different from bulk properties. The cyclic fatigue behavior of the bulk alloy is studied in detail.

Chapter 5 discusses the change in surface properties in the plastic zone accompanying a growing fatigue crack. It is shown that the plasticity index is very sensitive to local plasticity and the local changes indicate that ductility exhaustion is a precursor to fatigue crack propagation.

Chapter 6 discusses in detail the fatigue behavior of a polychloroprene rubber. Fatigue behavior is studied at different load ratios and strong indication of strain induced crystallization is observed, leading to life enhancement at high load ratios. A detailed fractographic examination is presented to illustrate the mechanisms.

Chapter 7 is a comprehensive review of the methods of determining fatigue crack growth resistance of metallic materials, including instrumented testing methods, spectrum fatigue testing, and standards for testing and analysis.

Chapter 8 treats friction and wear tests and gives a global and structured vision of the tribological tests existing from the classifications of tests according to the needs of the industry.

The final chapter presents a means of determining elastic properties by the nondestructive resonant vibrating method, developed for bulk and coated materials characterization.

It has been a pleasure realizing this collection, which will be very useful to research scholars, graduate students, and teachers.

The editor is grateful to all the 22 contributors to this book and to Pan Stanford Publishing for their help with the different processes of proofreading and editing.

N. Ranganathan
Tours, France
August 18, 2015

Chapter 1

Advanced Nanomechanical Test Techniques

Ben D. Beake,[a] Adrian J. Harris,[a] and Tomasz W. Liskiewicz[b]

[a] *Micro Materials Ltd., Willow House, Ellice Way, Yale Business Village, Wrexham, LL13 7YL, UK*

[b] *Institute of Engineering Thermofluids, Surfaces and Interfaces, School of Mechanical Engineering, University of Leeds, Woodhouse Lane, Leeds, LS2 9JT, UK*

ben@micromaterials.co.uk

1.1 Introduction

For many years the hardness of bulk materials and thick coatings has been determined by optical analysis of indentation marks. The development of thin coatings deposited by techniques such as physical vapor deposition (PVD) and chemical vapor deposition (CVD), to improve wear resistance, led to the requirement to measure their properties at a smaller scale. Initially such coatings were typically relatively thick (e.g., ~10 μm) and microhardness measurements could be performed to determine their hardness. However, as the thickness of the films reduced the reliable determination of their hardness by conventional optical means became impossible. Depth-sensing indentation (DSI) instruments have been developed to

Materials Characterization: Modern Methods and Applications
Edited by Narayanaswami Ranganathan

ISBN 978-981-4613-06-4 (Hardcover), 978-981-4613-07-1 (eBook)
www.panstanford.com

address this need and have become increasingly popular. The test technique is also called instrumented indentation testing (IIT) or nanoindentation and has progressed sufficiently for standardization to be required with the first international standard for DSI being released in 2002 and is currently in revision [1]. Provided instruments are well calibrated the data from nanoindentation tests are routinely analyzed by well-established contact mechanics treatments to provide the reduced elastic modulus and the hardness (or more strictly the mean contact pressure) of the test sample [2–3]. Conversion between nanoindentation hardness and Vickers hardness requires a little care. In addition to knowledge of the indenter geometry the actual contact areas used in the two definitions of hardness are slightly different necessitating the need for a geometric correction factor.

Over the last 25 years commercial nanoindentation test instruments (also called nanoindenters) have improved their resolution, their calibrations, and their ability to very precisely position where the indentations are made to obtain highly localized and accurate mechanical property information. Additionally, they have expanded the range of test techniques beyond simple nanoindentation, with several including some capability for nanotribological measurements (e.g., nanoscratch and nanowear testing), which has consequently greatly expanded their range of applications. There is a range of commercial nanoindenter designs, including electrostatic or capacitive actuation, and vertical or horizontal loading configurations. The design of one popular commercial test instrument, the NanoTest system from Micro Materials, combines electrostatic actuation with horizontal loading and an open test platform that has enabled its further development into a true multifunctional nanomechanical/nanotribological test instrument where tests can be performed with a range of contact geometries (Fig. 1.1) and environmental conditions (Fig. 1.2 illustrates the temperature and strain rate test envelope) [4]. The various tests provide complementary information and the data obtained can often more usefully map onto the actual conditions that the materials experience in use. It is becoming possible to move beyond basic characterization to the development of increasingly accurate prediction of the surface behavior.

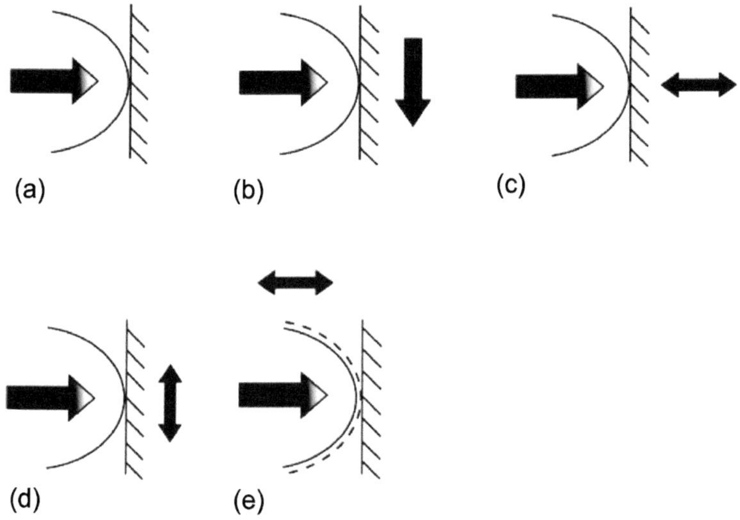

Figure 1.1 Range of nanomechanical testing capability in a commercial instrument (NanoTest system) illustrated for a spherical test probe. (a) Indentation contact, (b) scratch, (c) contact fatigue or impact by sample oscillation, (d) nanofretting or reciprocating wear, and (e) nanoimpact by probe impulse.

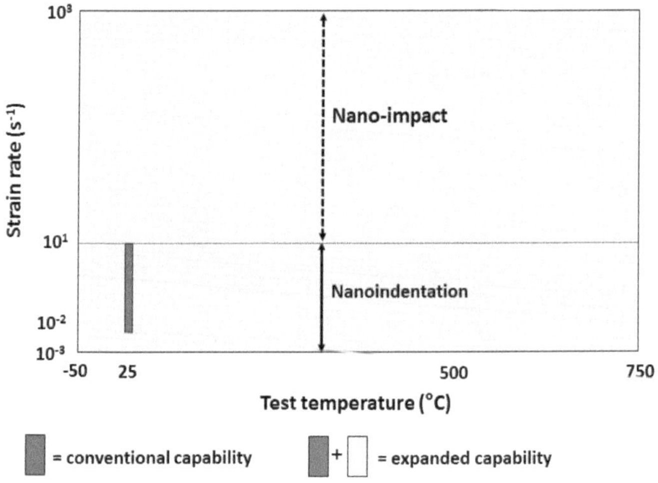

Figure 1.2 Temperature and strain rate test capability in a commercial instrument (NanoTest system).

1.2 Nanoindentation

1.2.1 *Contact Mechanics Theory*

Illustrative nanoindentation curves on fused silica, single-crystal tungsten and sapphire (0001) with a sharp pyramidal Berkovich diamond indenter are shown in Fig. 1.3. For a given indenter geometry the load–displacement curve recorded in a nanoindentation test can be thought of a "fingerprint" for a material as it contains information about the elastic and plastic properties of the sample under test.

The slope of the unloading curve at any point is called the contact stiffness. In this analysis, the reduced modulus, E_r, is calculated from the stiffness at the onset of the unloading, S, and the projected area of contact between the probe and the material, A_c, as

$$E_r = \frac{\sqrt{\pi}}{2\beta} \cdot \frac{S}{\sqrt{A_c}} \tag{1.1}$$

where β is the correction factor for the shape of the indenter. There is some ongoing debate over the exact value, though β is commonly taken as 1.034 for the Berkovich indenter geometry. The reduced indentation modulus, E_r, is directly determined in a nanoindentation test (also referred to as E'). However, the plane

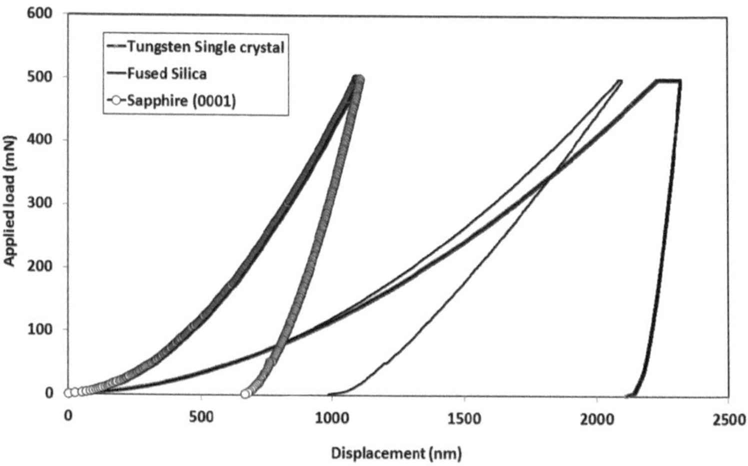

Figure 1.3 Nanoindentation curves on fused silica, single-crystal tungsten and sapphire (0001).

strain modulus, $E^* = E/(1 - \nu^2)$, can also be quoted (e.g., in ISO 14577 [1]) and when reporting nanoindentation data it is therefore necessary to specify clearly which modulus is being reported (E_r, E^*, or E). Conversion of the directly measured reduced modulus to the elastic (Young's) modulus of the sample requires that its Poisson's ratio be known or be reliably estimated. As elastic displacements occur both in the specimen and in the indenter (the indenter is not completely rigid), the elastic modulus of the sample is calculated from E_r using

$$\frac{1}{E_r} = \frac{1 - \nu^2}{E} + \frac{1 - \nu_i^2}{E_i} \tag{1.2}$$

where E and E_i, and ν and ν_i are the elastic modulus and the Poisson ratio of the tested material and the indenter, respectively. For diamond indenters, E_i and ν_i are 1141 GPa and 0.07, respectively. When a diamond indenter is used (i) on fused silica with $\nu = 0.17$, a reduced modulus of 69.6 GPa gives $E = 72$ GPa and (ii) for ceramics and hard coatings ν is typically 0.2–0.25. For sapphire with $\nu = 0.235$, an E_r of 314 GPa is equivalent to $E = 410$ GPa. (iii) For tungsten with a Poisson ratio = 0.28, $E_r = 320$ GPa converts to $E = 409$ GPa. The mean pressure or hardness (H) can be calculated as

$$H = P/A_c \tag{1.3}$$

where P is the applied load. Nanoindentation hardness is defined as the load divided by the projected contact area. However, in the definition of Vickers hardness the actual rather than projected area is used, resulting in a geometric scaling factor of 0.927 [5] and after correction for units a final relation of

$$H_V = 0.094545 H_{IT} \tag{1.4}$$

where H_V = Vickers hardness and H_{IT} is the DSI hardness.

1.2.2 *Practical Considerations*

1.2.2.1 Reference materials for calibration

Fused silica has proved the most popular material for calibrating nanoindentation instruments and the test probes (indenters) they use. It is inexpensive, highly polished, and mechanically homogeneous and has relatively little time-dependent behavior due to its high melting point (~1650°C). Typically its mechanical properties are quoted as $H \approx 9.2$ GPa, $E = (72 +/- 1)$ GPa and Poisson ratio $= 0.17$. In practice actual values may vary slightly from different suppliers. For example, a recent certified reference sample obtained from the UK National Physical Laboratory was supplied with $E = 73$ GPa and Poisson ratio $= 0.16$.

Despite being a default a reference sample popularly used for both area function and frame compliance calibration, its mechanical behavior in indentation is actually complex. Kermouche et al. noted that whilst normal glasses exhibit little densification on indentation, anomalous glasses exhibit large densification and different fracture behavior, and fused silica, which is the most anomalous silicate glass of industrial interest, can achieve 20% densification after indentation [6]. Its behavior is more complex than is generally supposed since (i) its near surface properties may be reduced by hydration, (ii) it is an anomalous glass whose elastic modulus increases with temperature [7], (iii) contact area determination can be influenced by radial dilation due to its high H/E [8], (iv) densification occurs under the indenter, (v) cracking occurs at high load, dependent on the acuity of the indenter [9], and (vi) by having a very high H/E ratio its sensitivity to small errors in frame stiffness was rather low.

Calibrations of test instruments can be more robust if supplemented by measurements into a metallic reference material with lower H/E such as aluminum or tungsten. ISO 145577 recommends the use of two materials of dissimilar E. An example of a successful calibration is shown below (Table 1.1) [10].

1.2.2.2 Choice of loading ramp

The standard contact mechanics treatment of the unloading curve data used to determine H and E is essentially based on elastic

Table 1.1 Multiple reference sample calibration (after [10])

Reference sample	E (experimental) (GPa)	E (expected) (GPa)
Fused silica	71.8 ± 0.4	72
Tungsten	417.5 ± 4.6	410
Steel	220.5 ± 1.5	220

contact mechanics with no provision for time dependency. It is important that time-dependent behavior does not influence the accuracy of the elastic modulus so the experimental conditions (load history) should be modified accordingly. Chudoba and Richter have detailed the problems that can occur when the holding period at peak load is too short [11]. For soft metals, the combination of slow loading, a long hold at peak load, and a fast unload ensures that the contribution of nonelastic behavior during unloading is minimized, or negligible in most cases. An alternative approach currently beginning to find favor is to develop more complex analytical treatments that can handle the time dependency.

1.2.2.3 Instrumental (thermal) drift

Since nanoindenters measure very small displacements high accuracy is required, that is, data should not be influenced by any thermal or mechanical drift during the test duration. For less thermally stable instrument designs, a practical solution has been to indent as fast as possible, although as mentioned above this fast loading and low hold time at peak load may result in some anelastic deformation during unloading for highly creeping materials. For all instrument designs it is important that when holding for a thermal drift correction (typically on the surface under the contact load immediately prior to the loading ramp and/or at 90% unloading) deformation due to creep or creep recovery not be erroneously thought to be due to thermal drift. An excellent example of this is the continuing extensive creep recovery that occurs on all polymeric materials. For highly stable nanoindenters this can be × 100 greater than the real thermal drift so mistakenly correcting data for it introduces error. Similarly if the sample is very soft (e.g., solder) then the initial load for thermal drift should not be used as deformation during this period is likely to be dominated by creep rather than drift.

1.2.2.4 The international standard for depth-sensing indentation: ISO 14577 parts 1–4

The International Organization for Standardization (ISO) has released an international standard ISO 14577 parts 1–3 entitled

"Metallic Materials: Instrumented Indentation Test for Hardness and Materials Parameters" in 2002, with part 1, "Test Method"; part 2, "Verification and Calibration of Machines"; and part 3, "Calibration of Reference Blocks." The recent extension to the ISO indentation standard (ISO 14577 part 4, "Test Method for Metallic and Non-Metallic Coatings") is also concerned with the experimental best practice for the determination of accurate hardness and elastic modulus of coatings independently of the influence of substrate. This is discussed further below.

Continuing time-dependent deformation during unloading affects the accuracy of the analysis of the unloading curve that assumes elastic behavior. In general a reliable strategy for minimizing/eliminating any influence of time dependency is a combination of (1) slow loading, (2) suitable hold period at peak load, and (3) fast unloading. This is illustrated schematically in Fig. 1.4a. During the hold period the creep rate decreases sufficiently that the continuing anelastic deformation during the faster unload is minimal, greatly increasing the accuracy of the elastic assumption in the unloading analysis. Figure 1.4b shows continuing creep during a 30 s hold at peak load at 500 mN on single-crystal Al and Fig. 1.4c shows nanoindentation curves from 100 to 500 mN on single-crystal Al.

The suitability of elastic unloading analysis for polymeric materials that show pronounced viscoelastic behavior has been questioned for the same reasons. Nevertheless, a similar approach to that recommended above for metals has recently been applied to a range of commercial polymers by Young et al. at NPL [12]. The elastic modulus measurements showed reasonably good agreement with supplier-quoted values (presumably from bulk methods) and a new atomic force microscopy (AFM) technique using Hertzian contact mechanics. Dub et al. found nonconstant elastic modulus versus load for copper samples when following the ISO 14577 approach [13]. The complex nonlinear behavior found was considered to arise from limitations to this method. Whilst the ISO standard does have some deficiencies and simplifications, in this case it is not clear that at least part of the experimental trend observed might have been due to the load history employed (constant loading and unloading time, in conjunction with no obvious hold at peak load for creep).

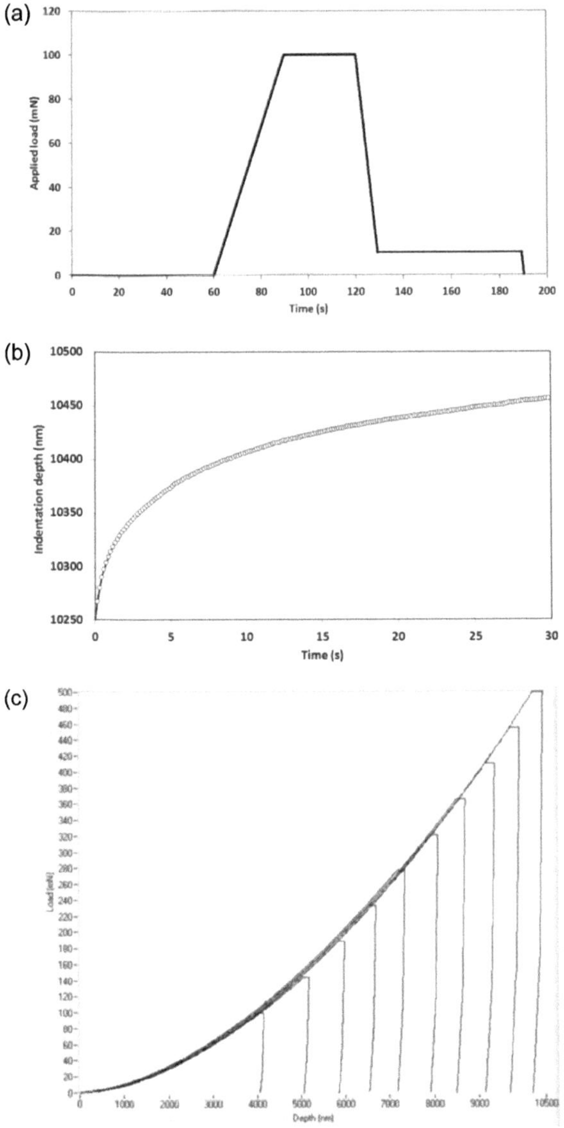

Figure 1.4 (a) Illustrative nanoindentation loading history for a metallic sample, (b) indentation creep during a 30 s hold at peak load at 500 mN on single-crystal Al, and (c) nanoindentation behavior at 100–500 mN on single-crystal Al.

1.2.2.5 Surface roughness

It can be beneficial to make measurements over a wide load range, and instrumentation with a restricted load range has limitations for (1) testing on rougher surfaces, (2) ensuring the accuracy of the extrapolation for coating-only modulus determination, and (3) investigating fracture behavior. For results to be completely free of surface roughness the ISO standard recommends that the indentation depth be $\times$ 20 greater than the surface roughness, R_a [1]. Therefore, provided the coating is sufficiently thick that the plastic stress field does not extend into the substrate, measurements at larger load will show smaller scatter due to roughness. If roughness is high then confidence can be increased by performing a much larger number of measurements. For arc-deposited PVD wear-resistant coatings designed for cutting tools the surface roughness can be typically $R_a = 0.1$ μm or more, requiring 30–40 indentations. The use of a calibrated contact load that can be increased for rougher samples (as in the NanoTest system) is an additional way to mitigate the effects of high roughness. Some nanoindenter designs use a reference ring method to subtract out some thermal drift but such approaches can prove problematic for rough samples and/or complex sample geometries. Schwarzer et al. commented that it was impossible to reliably test thick (Ti,Al)N and CrN coatings with such an instrument due to high surface roughness [14].

1.2.2.6 Determination of frame compliance

Removal of the contribution of elastic bending of the nanoindentation instrument from the raw data is required before analysis to determine H and E. Frame compliance is the inverse of stiffness. It is common to use compliance rather than stiffness as the relationship is simply

$$C_t = C_s + C_f \tag{1.5}$$

where C_t = total compliance, C_s = sample compliance, and C_f = frame compliance. Accurate determination of this frame compliance of the instrument is therefore essential, particularly for reliable modulus measurement. This can be done by direct or indirect methods. ISO 14577 describes both. The direct approach is less

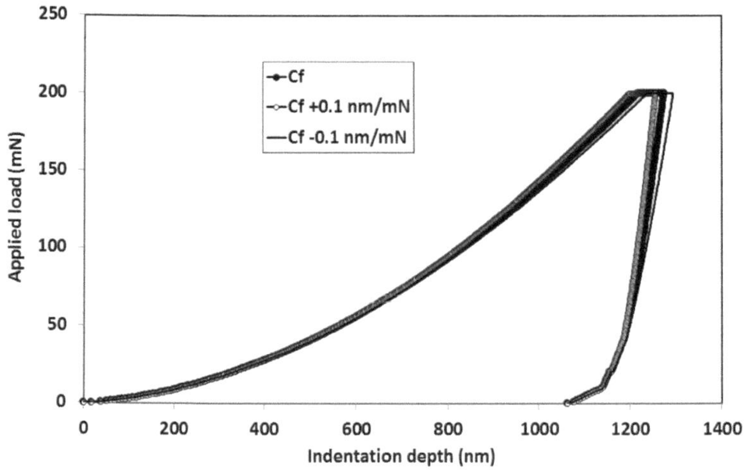

Figure 1.5 Indentation curve on single-crystal tungsten after correction of frame compliance, together with simulated curves with the frame compliance incorrectly determined.

well established but assumes nothing about the elastic properties of a reference material or the reliability of the contact mechanics and may prove increasingly popular in future. In principle a rigid contact is made between the load application part of the instrument and the rest of the frame so that all the measured deformation results from the compliance of the frame with a zero contribution from sample compliance [15]. The indirect method is much more common and relies on the sensitivity of the reference material(s) to the instrument stiffness and to the accuracy of the indenter area function used.

As an illustration of this, Fig. 1.5 shows the indentation curve on a tungsten reference sample analyzed with the correct frame compliance of 0.36 nm/mN for the instrument used together with simulated curves corrected for frame compliance of (0.36 + 0.1) nm/mN and (0.36 − 0.1) nm/mN rather than the correct value. The high sensitivity of the tungsten reference sample to the value of the frame compliance used in the calculation is due to its low H/E ratio. Since accurate frame compliance also requires an actual area function, the use of multiple reference samples is attractive. ISO 14577 recommends the use of two reference samples with widely differing elastic modulus. Accurate frame compliance calibration is even more important as the load increases into the microrange since the proportion of sample compliance to frame compliance decreases.

1.2.2.7 Determination of the area function

This can be done directly or indirectly. In common practice the area function is determined indirectly by indentations over a wide depth range into a homogeneous material with known elastic modulus (e.g., fused silica), with more measurements at the lower loads. A function is then generated, which relates the projected contact area and the contact depth so that a constant reference value of elastic modulus is returned over the fit range. The relationship between area and depth for an ideal Berkovich indenter is $A_c = 24.56h^2$. In practice all indenters are slightly rounded at the very apex and area functions are usually improved by the use of more measurements at lower load in the fit so that the relationship is accurate in this range. Area functions can be simple or apparently complicated relationships. On the simpler side, a function of the type $A_c = \mathrm{A}h + \mathrm{B}h^2$ is a reasonably robust approach that can often deliver modulus measurements to within about 2% of literature values across a wide depth range. It has the advantage of being able to be used to estimate the shape for depths above which calibration data exist so that measurements can be made on very soft materials. Nevertheless, not all pyramidal indenters are well fitted by such a simple relationship and polynomial fits (e.g., from the second order to the fifth or more) or spline fits can be preferable in many cases. However, higher-order polynomials should not be used outside of their fit range as they will diverge strongly. Another limitation of using fused silica is that above a threshold load (ISO 14577 suggests 75 mN for a Berkovich indenter but the exact value will depend on indenter sharpness) it cracks. Although such cracking is relatively minor for the Berkovich geometry compared to indenter geometries with more acute angles where discontinuities can clearly be seen in the loading curves, the cracks are clear in scanning electron microscopy (SEM) images of indentations and there will presumably be some subtle effect on the load–displacement curve even if obvious “pop-in” type features are absent.

The ISO also notes the possibility of more directly measuring the area function with a traceable (metrological) atomic force microscope. This has the potential advantage that any issues concerning the accuracy of the contact mechanics used in the analysis, such as radial dilation on fused silica [8], are eliminated, but the AFM measurement itself may be subject to some uncertainty, particularly at shallow depths. Dub et al. have suggested that AFM measurements are typically subject to up to 7% error [13]. Nevertheless, in future the AFM-based calibration may become more popular, particularly for certifying the geometry of master indenters for cross-calibration purposes. For day-to-day calibration the indirect method is expected to continue to remain popular as the most convenient way to assess any changes in geometry of an indenter, for example, after continued indentations into a hard tool coating or steel. In practice the area functions of new indenters can change significantly but often approach a quite stable, but not very sharp, geometry after a period of sustained usage.

1.2.2.8 Zero-point correction

Precise surface detection and assignment of the exact point at which the initial contact was made are important requirements for accurate hardness and elastic modulus measurements, particularly at shallow depths. There are different strategies for doing this, depending on whether precontact data are acquired or not. In one commercial instrument the zero point is defined by default at the position where the contact stiffness exceeds 200 N/m ($\times 2$ instrument spring stiffness). Moseson et al. have criticized this as being insufficiently reliable since they found it can lead to a false zero several hundred nanometers away from the actual point [16], although it is in widespread usage. The other main approach does not require precontact data, with the surface detected when a preset contact force is reached. In the NanoTest system, for example, the spring stiffness is typically $\times$ 10 lower than in other nanoindenters and the initial contact force can be varied over a wide range (e.g.,

0.5 – 50 μN), if required (minimized for accurate measurements in low-vibration environments, or significantly higher to combat high roughness and/or vibration levels). Since there is always some elastic deformation occurring before this point it is then necessary to correct the displacement data for the missing depth, either assuming Hertzian (since even sharp indenters always have some rounding at the tip) elastic loading or more generally assuming power-law or linear extrapolation, as suggested in the ISO standard. An example of a Hertzian correction is given by Eq. 1.6 [5].

$$F = C(h - h^0)^{1.5} \tag{1.6}$$

When indenting into a bulk material with well-known elastic properties it is therefore possible to back-calculate the radius of curvature at the tip, provided the data used are only to a small fraction of the radius using Eq. 1.7. Chiang et al. have noted that for the Berkovich the transition between the spherical end cap and the pyramidal shape occurs at h_c = indenter radius/4 [17].

$$F = \frac{4}{3} E \sqrt{R}\, h^{3/2} \tag{1.7}$$

1.2.3 *Indentation Size Effects*

The onset of nonelastic deformation and the associated critical load can be marked by a displacement excursion (a pop-in) or change in slope but this is not often the case, particularly for sharper indenters. Below this point the indentation behavior is fully elastic. It is often called the elastic–plastic transition and it marks the onset of dislocation motion in metallic materials, but in semiconductors pop-ins may be caused or complicated by phase transformation events occurring once a critical pressure is reached (e.g., pop-ins on Si usually occur when the contact pressure rises to about 12 GPa [18]). Sapphire is another material that can show well-defined pop-in behavior [19]. Dub et al. observed clear pop-ins of >50 nm on Cu (111) and Cu(100) at around 0.5 mN when using a 407 nm tip radius (blunt) Berkovich [13]. They reported mean pressure values at the moment of homogeneous nucleation of dislocations (i.e., at the pop-in load) for Cu (111) and Cu (100) surfaces of 16.8 and 9.3 GPa, respectively. Prior to the onset of plastic flow metallic materials

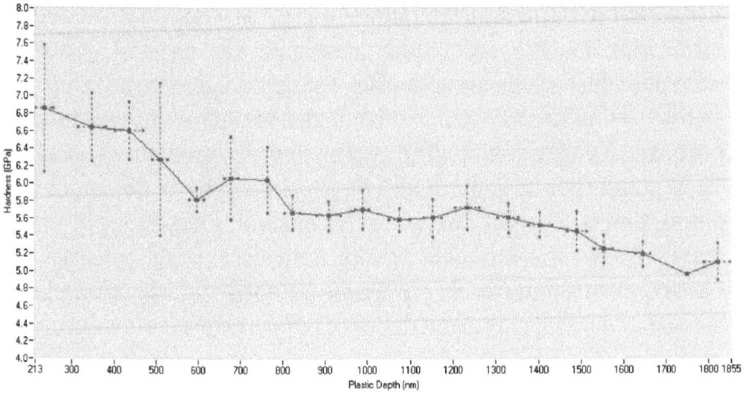

Figure 1.6 Polycrystalline tungsten shows a strong indentation size effect in hardness.

can support much higher contact pressure than their hardness postyield. Various indentation size effects (ISEs) can be observed, with hardness either increasing or decreasing as a function of depth. Several theories have been proposed to explain the most common ISE where hardness increases with decreasing indenter load, as in the example on tungsten shown in Fig. 1.6 (the modulus is constant over the same load range).

Arguably the most popular is gradient plasticity theory where geometrically necessary dislocations (GNDs) are generated under the indenter due to strain gradients. Decreasing the indentation size results in a greater density of GNDs and hence to higher measured hardness. The theory predicts that the total dislocation density within the plastically deformed volume varies as $1/h_c$, so $H^2 \alpha\ 1/h_c$. Nix and Gao fitted the hardness versus depth dependence with a two-parameter model, finding good agreement on a range of metals above indentation depths of 150 nm [20]. The two parameter model is shown in Eq. 1.8

$$H/H_0 = \surd[1 + (h^*/h)] \tag{1.8}$$

where H_0 and h^* are constants obtained by fitting. Starink and coworkers have recently shown improved agreement at low depths by taking into account the rounding of the indenter [21]. They were able to apply their model to predict (i) the hardness of UFG Al-1050, a material with previously unknown ISE, to below 150 nm and (ii) dislocation densities from indentation data, finding good agreement with published dislocation density data for single-crystal copper. In an interesting and as yet unvalidated idea Milman et al. have recently proposed a phenomenological approach to the problem of ISEs [22]. They noted that in general an empirical power law dependence is found; that is, $P \approx h^m$, where $m \approx$ constant. Plotting log P versus log h provides the exponent m. Alternatively, $H = Ah^i$ where A and i are constants and $i = m - 2$. These authors propose recalculating hardness of metals to a fixed depth of 1000 nm to enable comparison of studies at different loads.

1.2.4 *Choice of Indenter Geometry and Stress–Strain Curves*

Sharp indenters induce full plasticity at low load so that the mean pressure measured in the test is the hardness of the material. In general for a conical indenter the strain is proportional to the tangent of the cone angle and so is constant at different indentation depths (for a pyramidal indenter there is an equivalent angle). Blunt indenters, or spherical geometries, do not always generate full plasticity immediately and more force may be required. The strain induced by a spherical indenter is proportional to a/R and is a function of depth. This suggests the possibility to generate stress–strain curves by indentation, with the load partial unload technique being able to generate the entire stress–strain curve in a single test. The stress–strain curve in a tensile test can provide the following information for structural design: (i) Young's modulus, (ii) yield stress, (iii) work hardening exponent, (iv) ultimate tensile stress, and (v) strain at failure. Tensile tests suffer from limitations in specimen size and reproducibility and the possibility to obtain equivalent information from indentation is being actively investigated by several research groups. With finite element analysis (FEA) it is possible to generate the indentation curve from tensile stress–strain data but it is more challenging to uniquely solve the inverse problem and generate a tensile stress–strain curve from indentation data. Three main methods are (i) representative stress and strain, (ii) inverse analysis by FEA, and (iii) neural networks. Residual stress is an added complexity, with the indentation force required to reach a given depth under tensile stress being lower than in the stress-free state and under compressive stress the force required is greater.

1.2.5 *Plasticity Index and H/E*

Plasticity in contact is governed by the roughness and elastoplastic properties of the contacting surfaces. In an indentation contact plasticity is directly correlated with the ratio of hardness (H) and elastic modulus (E) [23, 24]. For a given indenter geometry the load–displacement curve recorded in a nanoindentation test can be thought of a "fingerprint" for a material as it contains information about the elastic and plastic properties of the sample under test. A useful dimensionless property is the plasticity index (PI), which is the plastic work done during indentation (W_p) divided by the total elastic (W_e) and plastic work done during the indentation, over a wide range of H/E values, with the relationship taking the apparent form

$$\mathrm{PI} = W_p/(W_p + W_e) = 1 - x(H/E_r) \qquad (1.9)$$

where x is a constant and E_r is the reduced indentation modulus. Prior to development of nanoindentation, a PI was proposed that related the deformation in rough contacts to the ratio of E_r/H multiplied by a geometric factor [25]. This original PI is a general surface texture parameter that combines material and topographic properties of solids in contact. The equation originally developed by Greenwood and Williamson [25] is shown in Eq. 1.10.

$$\psi = (E_r/H)\surd(\sigma/\beta) \qquad (1.10)$$

where $\psi = \mathrm{PI}$, $\sigma =$ standard deviation of the height of the contacting asperities, and $\beta =$ their average radius. Plastic deformation of asperities occurs even at minimal contact pressure when $\psi >> 1$ but when $\psi < 0.6$ deformation is largely elastic and plastic contact can only be caused if surfaces are forced together under very large nominal pressure. Although Eqs. 1.9 and 1.10 differ in form both these plasticity indices reflect essentially the same behavior with varying mechanical properties, that is, increasing H/E results in a lower PI and more elastic contact [26]. Provided H/E_r is not greater than about 0.1 a good approximation to experimental data is given by Eq. 1.9, although there is significant nonlinearity above that and Chudoba and Jennett have noted that it should not be possible to produce plasticity with a Berkovich beyond $H/E = 0.18$ [8]. FEA gives $x \approx 5$ [15]. Beake et al. have reported that with several different Berkovich indenters values around 6–7 were found over a wide range of indentation peak forces for coatings and bulk materials such as copper, aluminum, tungsten, bearing steel, stainless steel, and WC-Co, whilst only fused silica and K7 glass show lower x values very close to those predicted by Cheng and Cheng [26, 27]. H/E is well correlated with how elastic the contact is in other related mechanical contact situations such as scratching/sliding [28–30] or impact/erosion [31, 32]. Higher values result in higher critical loads for the onset of yield (nonelastic deformation) in indentation or scratch testing.

1.2.6 *Hardness and Elastic Modulus of Coatings*

An often-quoted rule of thumb is the so-called 1/10 rule, where the relative indentation depth (RID; indentation depth/coating thickness) should be less than 0.1 to determine coating-only properties independently of the substrate [1, 5]. In practice this usually works well for hardness but less well for elastic modulus. The ISO standard recommends Berkovich indenters be sharper than 250 nm end radius for hardness measurement. Whilst this may not be really necessary for thicker coatings, sharper indenters are preferable for hardness measurements on very thin films. Accurate hardness measurement of a coating relies on development of a fully plastic zone within the coating. If the applied load is too small then the results can be influenced by surface roughness and the plastic zone may not be fully developed. If the applied load is too large then the plastic stress field may extend beyond the coating–substrate boundary and the measured response is therefore a composite of the plastic properties of both the coating and the substrate. Theoretical models based on approaches such as weighted area or volume law of mixtures have been developed to estimate the coating-only hardness from this composite response, with later work focusing on energy-based (work of indentation) models that have been adapted to include the effects of both plasticity and cracking [33]. As mentioned above, the elastic modulus measured at RID = 0.1 can be a poor estimate of the coating modulus, especially when there is a large elastic mismatch between coating and substrate. For example, Korsunsky and Constantinescu have predicted that when a coated system of $E_c = 500$ GPa, $E_s = 100$ GPa, $\nu_c = 0.2$, and $\nu_s = 0.3$ (i.e., approximating to a stiff ceramic coating deposited on a much less stiff metallic substrate) is indented with a conical indenter the modulus is 25% lower at an RID of 0.1 than the true coating value [33]. ISO 14577 part 4 recommends making measurements over a range of indentation depths, provided that a/t_c is less than 2.5, and extrapolating to zero depth to obtain a value of the coating modulus independent of the substrate. In practice a linear extrapolation is a fairly robust approach, which usually yields improved values from those determined from measurements to an RID of 0.1, although measurements to an RID of 0.03–0.05 can also work well.

1.2.7 *Ultrahard Coatings*

Accurate and traceable calibration by methods such as ISO 14577 can help validate or question the mechanical properties of new coatings. Veprek et al. have claimed deposition of ultrahard nanocomposite coatings with hardness of $\geq$100 GPa (i.e., higher than diamond) [34]. This highly contentious claim has been the subject of an excellent critical review by Fischer-Cripps, Bull, and Schwarzer [35]. These authors have provided compelling evidence for these reported hardness values to be overestimated by a factor of ~2. They were able to show that the analysis approach taken by Veprek et al. that obtained the correct hardness of a ultrahard sample of known properties (industrial diamond) nevertheless underestimated the elastic modulus by a factor of ~3, proving that the calibration and analysis method was not accurate. Additionally, since the elastic modulus of the nanocomposite coatings in question was only in the region of 450 GPa, it should theoretically not be possible to produce coatings with hardness over ~65 GPa due to the limit in H/E with a conical indenter.

1.2.8 *Depth Profiling*

There are three popular methods for obtaining properties such as hardness, modulus, and contact stiffness as a function of depth. These are (i) indentation to differing peak loads, (ii) dynamic indentation (otherwise known as continuous stiffness measurement, or CSM), and (iii) multicycle loading (or load–partial unload testing). The most obvious, although quite time consuming, is simply to perform indentations at different locations with different peak loads and consequently stress fields. The CSM method has recently been called into question by Cordill et al., who noted that the oscillation used with dynamic indentation techniques alters the measured hardness value of ductile metallic materials, especially at depths less than 200 nm [36]. The alteration in the hardness is due to the added energy associated with the oscillation that assists dislocation nucleation. In other words, there is inherent risk in trying to determine properties by oscillation as you are completely changing the dislocation density of the material you are trying to measure. However, provided the test instrument has good thermal stability nanoindentation experiments can easily be performed over longer duration as multicycle tests, enabling effective hardness and modulus determination as a function of depth without the risk of altering the surface of soft materials by imposing an oscillation. Figure 1.7 shows an illustrative load–partial unload test result for a hard diamond-like carbon (DLC) film deposited on silicon. The inflexion in the curve marked by an arrow occurs at the point at which the plastic stress field is no longer completely within the coating and the measured hardness becomes a composite response of the coating and the softer substrate.

1.2.9 *Mapping and Positioning*

Nanoindentation instruments have precision stages enabling accurate positioning in discrete phases to obtain highly localized mechanical property information. Experiments are typically performed as large automated arrays (e.g., 10 × 10 or 20 × 20 grids of indentations) or more focused tests involving a smaller number of indentations precisely positioned in the phases of interest.

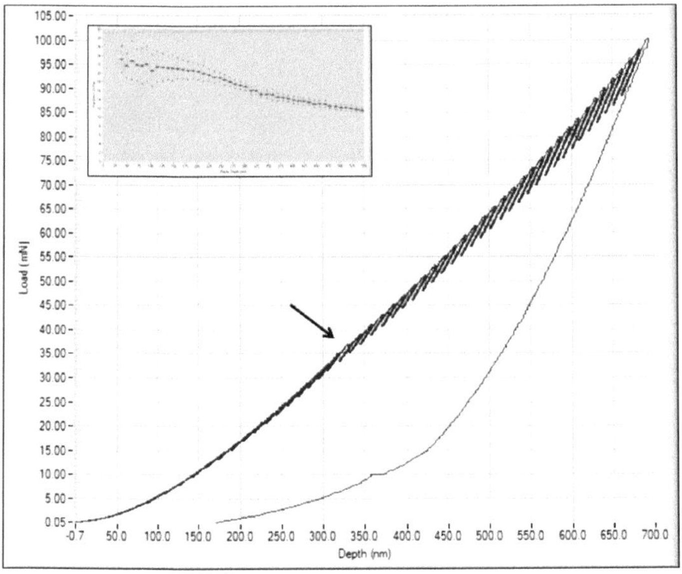

Figure 1.7 A 50-cycle load–partial unload test on a 2 μm DLC film on Si. The inset shows variation in hardness with depth from repeat tests.

An example of grid mapping is shown in Fig. 1.8. Here 400 indentations have been made into a Ti6Al4V aerospace alloy as a 20 × 20 grid, with adjacent indentations spaced 20 μm apart. Average values of hardness and reduced elastic modulus were (4.5 ± 0.5) GPa

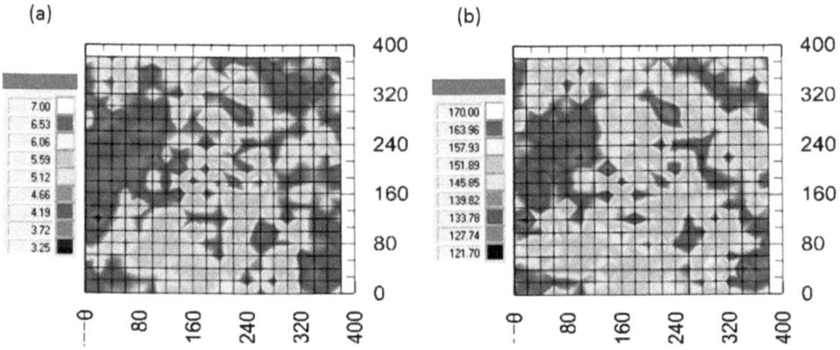

Figure 1.8 A grid of 400 indentations into Ti6Al4V reveals subtle mechanical property variations across the surface (a) hardness map and (b) reduced elastic modulus map.

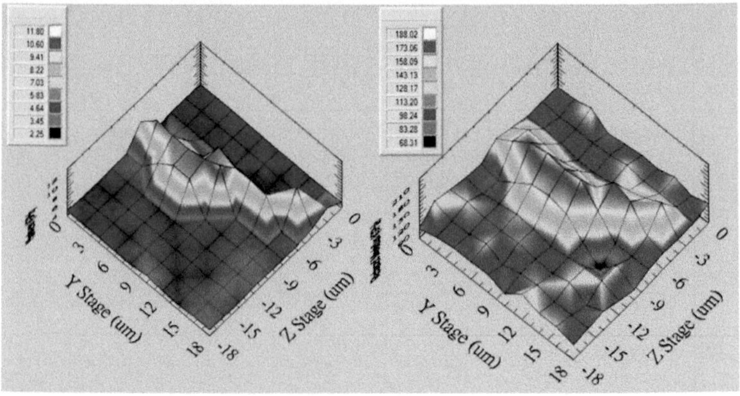

Figure 1.9 Mapping the mechanical properties of an intermetallic inclusion in a 7000 series Al alloy.

and (139.7 ± 7.5) GPa, respectively. More information is revealed in the hardness and modulus maps, with distinct areas of lower mechanical properties.

When performed to much lower load (and therefore depth) indentations can be placed closer together (e.g., 1–2 μm pitch) to obtain higher-resolution maps. In Fig. 1.9 inclusions of an Al_7Cu_2Fe intermetallic phase can be seen to have completely different mechanical properties to the bulk matrix.

1.2.10 *Fracture Toughness*

Depending on the load, material, indenter geometry, and environment, five major types of cracks occur in the indentation of brittle bulk materials or thick coatings [37]. These are (a) cone cracks, (b) palmqvist radial cracks, (c) median cracks, (d) lateral cracks, and (e) half-penny radial cracks. Fracture toughness can be determined from the crack systems by either stress-analysis-based or energy-based models. The application of these to the determination of the fracture toughness of thin coatings has been well summarized in an excellent recent review by Chen [38]. Lawn and coworkers showed that for a well-developed radial/median crack system, that is, half-penny like cracking caused by indentation, the K_c fracture toughness of the material was proportional to the applied load multiplied by the average crack length c raised to the power of $-3/2$ [39] (Eq. 1.11).

$$K_c = C(E/H)^{1/2} P/c^{3/2} \tag{1.11}$$

where C is a geometry constant commonly taken as 0.016 for Berkovich or Vickers and 0.032 for cube corner indenter geometries. However, as the thickness of the coating is reduced it becomes more difficult to apply this equation due to the interference of the substrate and interface with the coating crack system. If the load is reduced to avoid this complication then the crack system is not sufficiently well developed to apply the analysis. For thinner coatings the energy-based models refined by Bull and Chen can be more applicable [40].

1.2.11 *Micropillar Compression, Microcantilever Bending, and Push-Out Tests*

Introducing an indenter into a sample surface produces a complex stress field, which can make determination of mechanical properties more challenging. A sharp indenter such as a Berkovich will drive plasticity almost immediately on contact. Whilst this is required in order to measure hardness it does present limitations when material behavior within the elastic limit is to be investigated. Additionally, as an indenter continues to penetrate a sample surface, an exponentially increasing volume of material becomes involved in the deformation. As the indentation progresses, there will constantly be new material entering progressively advanced stages of deformation and this can create difficulties in understanding the deformation mechanisms at work. In macroscale experiments such problems are obviated by using tensile or compressive load application to a sample of a known cross-sectional area. This produces a constant stress at any given point along the specimen (at least until plasticity is initiated locally). Modern sample preparation techniques, in particular focused ion beam (FIB) milling, now allow the manufacture of microscopic equivalents of the samples used in the conventional macroscale techniques. The most popular sample geometry used is the micropillar [41–43]. There is also growing interest in microcantilevers (or "bend bars"), which may be produced using the same technique.

1.2.11.1 Micropillar compression

For almost a decade now interest has been growing in the field of micropillar compression experiments. Uchic et al. used the technique to demonstrate size effects on the measured strength of Ni and three Ni-based alloys [41]. They demonstrated that micropillars of practically manageable dimensions (0.5 to 40 μm diameter) were sufficiently small to show the relationship between strength- and volume-based phenomena such as dislocation nucleation and movement. Once the pillars have been produced they are carefully positioned in front of a flat topped indenter commonly known as a flat punch. This is achieved by positioning the sample precisely in the crosshair of the nanoindenter's microscope and then moving to the indenter position (the vector from the microscope cross hair to the indenter center is calibratable and automatically executable on most nanoindenters to better than 1 μm precision). Once in position the load is applied to the micropillar using the flat punch indenter. Data are easily converted to engineering stress and engineering strain due to the dimensions of the micropillar being well characterized and there being no constraining material surrounding that being tested. Although pillars are the most common geometry for microcompression experiments, other structures have been tested. By compressing of variously shaped wall-like structures with differing length-to-width ratios Jennett et al. provided evidence that yield strength is determined by the smallest dimension of the structure ("thinness") [44]. Their results showed that yield stress was determined by wall width, with length playing little or no effect.

1.2.11.2 Microcantilevers

Microcantilevers are produced in a similar manner as micropillars. A FIB is used to cut out an elongated three-sided trough, leaving an island of material of the length and width of the required cantilever. The FIB is then reorientated such that it can be used to undercut the cantilevers, allowing them room to deform as the load is applied.

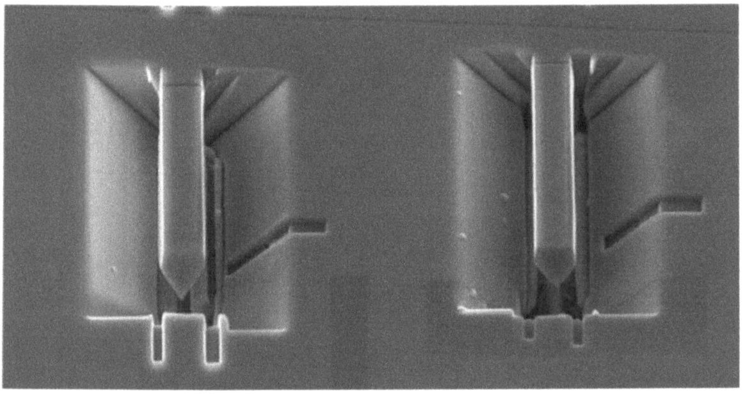

Figure 1.10 SEM image of two microcantilevers (4 μm wide by 20 μm long) cut into a silicon wafer. These samples have been prepared for brittle-to-ductile transition experiments over a range of temperatures and as such have had a crack initiation notch machined into them near their roots. Image courtesy of Dr. David Armstrong, University of Oxford.

Similarly to micropillars, microcantilevers may be targeted using an integrated optical microscope but more commonly a 3D imaging stage is used. This allows the cantilever to be imaged whilst mounted on the nanoindenter. The sample is mounted on a high-precision feedback-controlled piezoelectric stage. The sample is either raster- or bidirectionally scanned, whilst the indenter is held on the sample surface under a very low load (~1 μN). The scan parameters are defined by the user; these include the dimensions of the imaged area and the resolution of the image (number of scanned lines). A 3D image of the sample is produced from a combination of two sets of coordinates from the piezoelectric stage and the third set of coordinates from the indenter depth sensor.

The resultant image may then be used to select the point(s) at which the indenter is applied to the cantilever to apply the bending load. The distance between the root of the cantilever and the chosen contact point(s) is easily determined from the 3D image and as such the stiffness of the cantilever may be determined once the bend test has been completed. The stiffness value may then be used to determine the elastic modulus of the sample material using beam theory. This approach is well documented by Armstrong et al. [45],

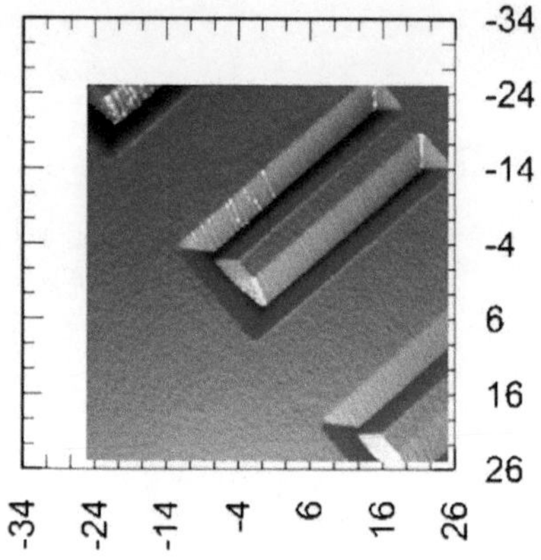

Figure 1.11 3D "nanopositioner" image of cantilevers machined into W. Note that the true cross section of the cantilevers is not seen due to the geometry of the Berkovich indenter used for imaging. Image size is 50 × 50 μm.

where cantilevers were FIB-machined into selected copper crystals whose orientations had been previously characterized by electron back-scatter diffraction (EBSD). The combination of the data from both techniques demonstrated the anisotropic nature of mechanical properties as a function of crystal orientation.

1.2.11.3 Fiber push-out

The most performance critical aspect of a fiber-reinforced composite material is the strength of the bond between the matrix and the reinforcing fibers. It is therefore desirable to be able to measure the strength of this interface. One technique for making such a measurement is the fiber push-out test. These tests are undertaken using a nanoindenter fitted with a flat punch indenter. The target fiber is selected under the instrument microscope and the sample moved to the indenter position such that the indenter contacts concentrically on the selected fiber (Figs. 1.12 and 1.13). Jouannigot et al. [46] have used this technique to mechanically characterize the fiber/matrix interface of carbon/carbon composites.

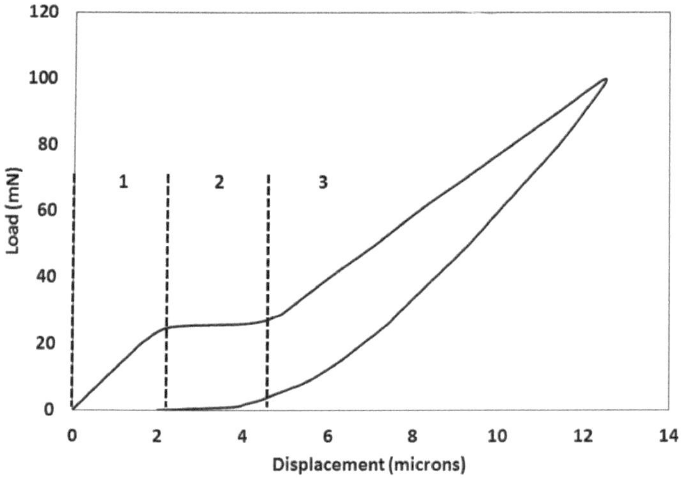

Figure 1.12 Load–displacement curve from a fiber push-out with a diamond punch.

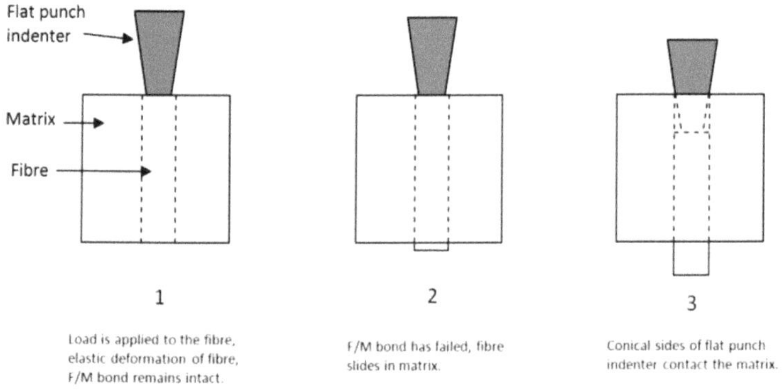

Figure 1.13 Movement of fiber during the loading curve.

1.3 Nanotribology (Nanoscratch and Wear)

Increasing use of thin films and coatings from a few nanometers to about 1 micron thickness for improved wear resistance in microelectromechanical systems (MEMS) and general thin-film applications requires that their mechanical properties and tribological performance be optimized. Since the test conditions in the conventional macroscale scratch test ($R = 200$ μm) were specifically chosen for fairly thick (≥3 μm) coatings they are not appropriate for much thinner films and nanoscratch and nanowear tests are needed. These tests typically use considerably lower load combined with a smaller probe radius than macroscale tests to achieve the required combination of surface (and lateral) sensitivity and high contact pressure to drive film failure.

Integrated precision control of the movement of the sample stage independently of the load application during the test enables various topographic, scratch, and wear tests to be performed. In the NanoTest system, for example, there are four distinct types of tests:

(i) Topographical scan, where the sample is scanned under a light (no wear) load to obtain a line profile of the surface and quantitative roughness statistics
(ii) Progressive (ramped) load scratch with a linear load ramp, either initially or after a set levelling distance
(iii) Topography-scratch-topography (three-scan) test
(iv) Multipass wear tests involving various combinations of topography and either ramped or constant-load scans

The tangential (frictional) force can be recorded simultaneously, which provides valuable confirmatory information on the location of failure relative to the moving test probe.

1.3.1 *Critical Load Sensitivity to Intrinsic and Extrinsic Factors*

Steinmann et al. showed that the critical load in the scratch test can be influenced by a range of extrinsic and intrinsic factors [47]. The influence of the scan parameters of loading rate and scan speed on the critical load has been investigated for a range of Si-doped DLC coatings on glass that show clear film failure in nanoscratch tests [48]. There was considerably less sensitivity to these scan parameters than has been observed in macroscale scratch testing. At the macroscale the $\mathrm{d}L/\mathrm{d}x$ ratio, where $L =$ normal load and x = scratch distance, was found to be an important parameter, with the critical load decreasing when $\mathrm{d}L/\mathrm{d}x$ decreases.

Steinmann et al. recommended tests be done with a fixed dL/dx of 10 N/mm, suggesting that when dL/dx decreases, the probability of encountering defective adhesion within a certain load range increases, resulting in a decreased critical load. This explanation may not be true for highly homogeneous coatings that do not fail by selective failure of poorly adhering regions and so exhibit very consistent L_c values, and also does not consider the possibility of cohesive failure.

In contrast at the nanoscale, no dependence on critical load was found for Si-doped DLC coatings on glass on either (1) scratch speed, (2) loading rate, or (3) increase in load per unit scratch distance (dL/dx) when dL/dx is much less than 1 N/mm, whilst critical loads only slightly increase at higher dL/dx (1–5 N/mm). More recently Beake et al. have investigated the variation in critical load for failure of a 80 nm tetrahedral amorphous carbon (ta-C) film on Si over a 100-fold range of dL/dx without finding any clear dependence [48]. These results suggest that nanoscratch tests under significantly different loading conditions can be compared directly.

1.3.2 *Topography-Progressive Load Scratch-Topography Multipass Scratch Tests*

An improvement on the basic nanoscratch test is the three-scan procedure (with residual scan) that enables identification of failure mechanisms, the role of stress in particular, in more detail. All the major nanoindenter manufacturers offer this as part of their nanoscratch testing capability. The first reported multipass test of this type was described in detail by Wu, and coworkers, from IBM in 1989 [50]. Although common thereafter, it was not until after it was realized that it was possible to remove instrument compliance from the measured deformation that true nanoscratch and nanowear depth data could be displayed [30]. The example below (fig. 1.14) shows similarity in the low load response of Si(100) to indentation and scratch testing with the same spherical probe.

By performing three-scan progressive load nanoscratch tests it is possible to determine the critical load for the onset of nonelastic deformation since this is the load at which the residual scratch depth is no longer zero. The mean pressure at this point can be used as an

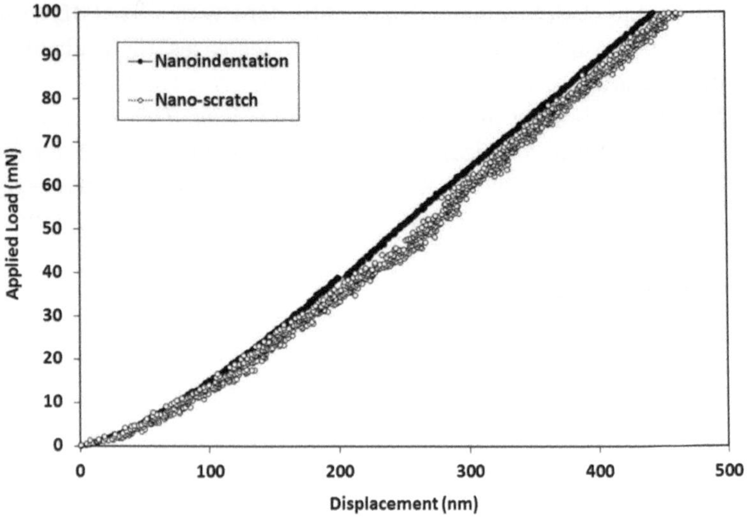

Figure 1.14 Load vs. depth curves in nanoindentation and nanoscratch tests on Si(100) with a 4.6 μm spherical probe.

estimate of the yield pressure. By increasing the load further it is possible to identify the critical loads for transitions for cracking and delamination events and converting these to mean pressures.

1.3.3 *Influence of Probe Radius and Geometry*

The choice of probe radius in the nanoscratch test should be considered carefully. Spherical probes with end radii of typically 1–10 μm are commonly used for nanoscratch testing. If probes are too blunt or the instrument used has a relatively small maximum load then it is not possible to scratch the coatings to failure. Spheroconical probes with a ~5 μm end radius represent an effective choice ensuring that coating failure is observed within the force range of the instrument but without risking rapid blunting of the indenter due to repeat scanning that can be observed when sharper probes are used (such as Berkovich indenters), especially when scratching hard and rough coatings. The use of the spherical indenters has the additional advantage that the contact pressure analysis described below can be directly applied to the nanoscratch data, provided the

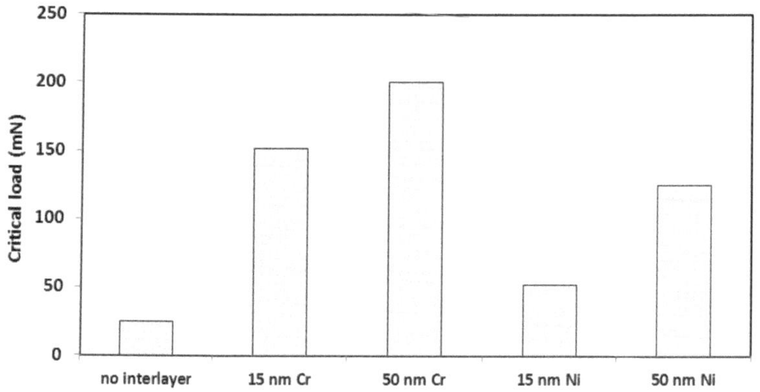

Figure 1.15 Influence of interlayer composition and thickness on critical load for Pt–Ir coatings.

instrument software has the capability to remove the contribution of instrument compliance, sample slope, and roughness from the measured deformation data so that true nanoscratch and nanowear depth data are displayed. Precalibration of the indenter radius by nanoindentation is necessary. Nevertheless, other geometries such as Berkovich [51] and cube corner have been used [52]. The latter combines a small end radius with high acuity and contact strain and so is particularly suited to the scratch testing of very thin films. Klocke et al. have used cube corner indenters to perform nanoscratch tests of 250 nm Pt–Ir coatings on cemented carbide tools used for precision glass molding [52]. They compared the scratch behavior of the Pt–Ir coating without interlayers and the same coating with 15 and 50 nm thick Ni or Cr adhesive interlayers. Without an interlayer the coating failed at 25 mN but the critical load was much higher (×2–×8) on the coatings with the thin interlayers (Fig. 1.15). For a given thickness the coatings with the Cr interlayer outperformed those with the Ni interlayer and for a given interlayer composition, the critical load for coatings with thicker interlayers was greater.

This appears to be a good example of dimensioning, where the probe sharpness was increased to improve the sensitivity to the adhesion at the interface. Schwarzer et al. have used advanced simulation tools to dimension macroscale scratch tests of thick multilayered cutting tool coatings [14]. By adjusting the scratch probe radius and applied load they were able to show the possibility to adjust the position of the depth of the maximum von Mises stress in multilayer coatings. By appropriate "tuning" they could position the maximum stress near the different interfaces in the coating–substrate system.

There have been occasional reports that the scratch depth under load may be greater for constant-load than in progressive-load testing but this does not appear to be the case in general. Beake et al. reported that for TiN on tool steel the deformation in constant-load and ramped-load tests was almost identical at the same load [53]. This implies that the critical load obtained in a progressive load scratch test is a reliable choice of determining a fractional subcritical load to assess the film resistance to low cycling fatigue in a nanowear test.

1.3.4 *Contact Pressure*

Recently the capability of the progressive load multipass scratch technique has been extended by the development of a novel analytical treatment to obtain accurate scratch depth data after correction of the raw data for the contribution from instrument compliance, sample topography, and sample slope [54]. The method described in Ref. [54] enables the yield stresses and the pressure required for the failure of the film to be estimated from contact mechanics, assuming the geometry of indentation, provided spherical indenters are used. By using spherical probes the contact depth (h_p) in an indentation contact is given by

$$h_\mathrm{p} = (h_\mathrm{t} + h_\mathrm{r})/2 \tag{1.12}$$

where h_p is the contact depth, h_t is the on-load scratch depth, and h_r is the residual depth from the final scan. The contact radius (a) is determined from Eq. 1.13, where R is the indenter radius.

$$a = \surd(2Rh_\mathrm{p} - h_\mathrm{p}^2) \tag{1.13}$$

$$P_\mathrm{m} = L/\pi a^2 \tag{1.14}$$

The contact pressure, P_m, at any point along the scratch track is given by Eq. 1.14, where L is the applied load. To apply this approach to the nanoscratch data it is necessary to assume that:

(1) the presence of a tangential load does not influence the pressure distribution too greatly (i.e., the measured friction coefficient is well below 0.3);
(2) the radius of the indenter is constant;
(3) the sliding speed is sufficiently slow and contact sufficiently close to elastic that the load is supported on the rear of the indenter; and
(4) the indenter can reach the bottom of the scratch track in the final topographic scan.

In practice it appears that these conditions can often be met in the nanoscratch test [54, 55], although the approximation becomes less valid as the geometry moves away from Hertzian. The methodology has been validated for thin films on silicon wafers with good agreement between (i) scratch hardness independently determined from optical measurements of scratch widths and (ii) contact pressures for film and substrate yield events.

1.3.5 *Influence of Film Stress and Thickness*

Beake et al. have reported nanomechanical and nanotribological (nanoscratch and nanowear) characterization of a wide range of thin films on Si, with thicknesses from 5 nm (ta-C) to 1500 nm (TiN), performing nanoindentation with Berkovich indenters combined with nanoscratch testing with spherical indenters [49, 54–57]. A key motivation for studying several of these was to understand the interplay between film thickness and interfacial toughness for MEMS and protective thin-film applications. The reliability of Si-based MEMS devices is limited by stiction forces when contact occurs. Applying very thin, low surface energy and low stress coatings can alleviate this problem but it is critical that they be deposited optimally.

The film thickness is a key parameter influencing the critical load in the nanoscratch test. In principle, film thickness can have two opposing effects: (1) Thicker films that are harder than the underlying substrate provide more load support and so delay the onset of the substrate deformation that is often the precursor of film failure (higher critical load, and (2) thicker films can be more highly

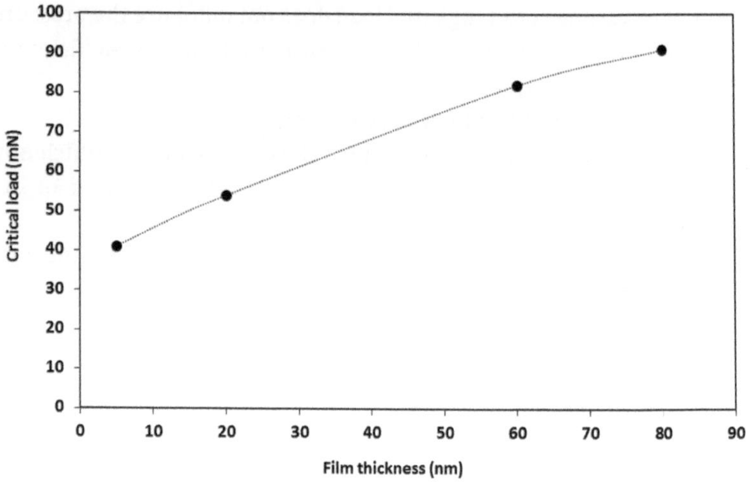

Figure 1.16 Variation in the critical load of 5–80 nm ta-C films on Si with thickness.

stressed and more easily through-thickness crack and delaminate when deformed (lower critical load) since the driving force for spallation to reduce stored elastic energy is greater. The relationship between film thickness and critical load can be quite complex in practice.

Nanoscratch studies on 200–1000 nm closed-field unbalanced magnetron sputtering (CFUBMS) a-C [57], 150–600 nm plasma-enhanced chemical-vapor-deposited (PECVD) Si:a-C:H [48], and 5–80 nm ta-C [56] have reported that the critical load for total film failure in the nanoscratch test can be a strong function (often close to proportional) with film thickness reflecting enhanced load support and substrate protection. Figure 1.16 illustrates the dependence for ta-C films when using a 3.1 μm spherical probe. Provided the films are not too stressed it has been reported that the ratio L_c/t_f is approximately constant.

Shi et al. reported nanoscratch data for 200 nm and 1000 nm a-C films deposited with varying substrate bias voltage (−20 to −140 V) by CFUBMS [57]. They found that on wear resistance grounds a suitable strategy for optimizing wear resistance for MEMS applications was to maximize H/E. For 200 nm films the highest

H/E materials performed best since the intrinsic stress in the film was relatively low (no recovery spallation and low H/E). For ~1 μm a-C films they observed a general relationship between H/E and the scratch test critical loads. When H/E_r is ~0.09–0.1 there was a tendency of the films to delaminate behind the moving probe at low critical load. Beake et al. previously reported a similar correlation for ~0.8 μm Ti–Si–N nanocomposites films on Si [50]. Higher H/E led to higher critical loads for elastic–plastic transition and also for the total film failure occurring in front of the probe. However, when H/E_r was > 0.09 film failure was always by tensile failure behind the probe (also described as unloading failure). The high tensile stress behind the probe in the nanoscale scratch test can lead to complete failure for hard films on silicon. More information can be provided by nanowear tests.

1.3.6 *Repetitive Nanowear*

Constant-load, unidirectional multipass scratch testing was first described by Bull and Rickerby in 1989 [58] and has been shown to be an effective low-cycle fatigue test. The same approach can be applied to micro- and nanoscratch testing. Constant-load nanowear tests are often used to determine rates of sliding/abrasive wear and investigate the role of fatigue. The low-cycle nanowear experiments can often be much more informative regarding the influence of thin-film stress leading to poor adhesion than single-scratch tests. When compared to progressive-load nanoscratch testing, nanowear testing has the advantage that the load can be varied to tune the maximum von Mises stress to be close to the coating–substrate interface. The friction force is sensitive to the onset of coating failure.

Shi et al. performed subcritical-load nanowear testing to investigate the performance of 1 μm a-C films that fail in single nanoscratches at ~200 mN [57]. The wear test loads were chosen so that the maximum stresses were either within the film (at 50 mN) or within the substrate (at 150 mN) so that the additional stress due to the imposed stress field from the scratch reaches the interface for the higher-load test. In the lower-load wear test the contact is almost completely elastic with residual wear depths under 100 nm. An approximately inverse correlation between H/E and the residual depth was found, consistent with the work of Matthews and Leyland, who have shown that increasing H/E reduces plasticity [28, 29]. An increase in residual wear depth and decrease in scratch recovery with each wear cycle confirm that the low-load multipass wear test is a fatigue process. In marked contrast, the films failed after a few cycles at 150 mN. As $150/L_u = 0.7$ this is perhaps not too surprising, particularly as the load is great than that at which initial edge cracks form (L_{c1}). Nanowear under these conditions is a very low-cycle fatigue test where plasticity and microfracture dominate and the harder films deposited under high bias perform poorly. The behavior of 1 μm a-C films with a very high ratio of H/E, deposited under high substrate bias, is very strongly dependent on the test conditions, performing well at low-load wear but very poorly in more highly loaded contact. They exhibit low critical loads in progressive-load scratch tests and when they fail in the nanowear test there is extensive delamination outside of the scratch track. The combination of nanowear tests at different loading levels provides information regarding the suitability of the a-C films for contact applications.

Repetitive scratch tests can also be informative at the microscale, typically using 25 μm end radius diamond probes in conjunction with applied load in the range 0.5–5 N. Beake and Ranganathan used this approach to investigate the low-cycle fatigue behavior of mono-, bi-, and trilayer hard coatings deposited on WC-Co by CVD [59]. They found a strong influence of the total coating thickness on the microscratch and microwear behavior. The microwear behavior of the trilayer coating system (3 μm TiN on 4 μm Al_2O_3 on 5 μm TiCN) is shown to be superior to the bilayer and monolayer coatings in the figure below (Fig. 1.17).

1.3.7 *Surface Roughness*

Beake et al. have recently investigated the influence of coating thickness, roughness, and direction of scratching relative to grinding marks on the behavior of TiN coatings on M42 steel in single- and multipass nanoscratch (nanowear) tests [53]. TiN coatings were deposited on M42 steel as this is a model system displaying clear failure with comparative frictional data available from previous

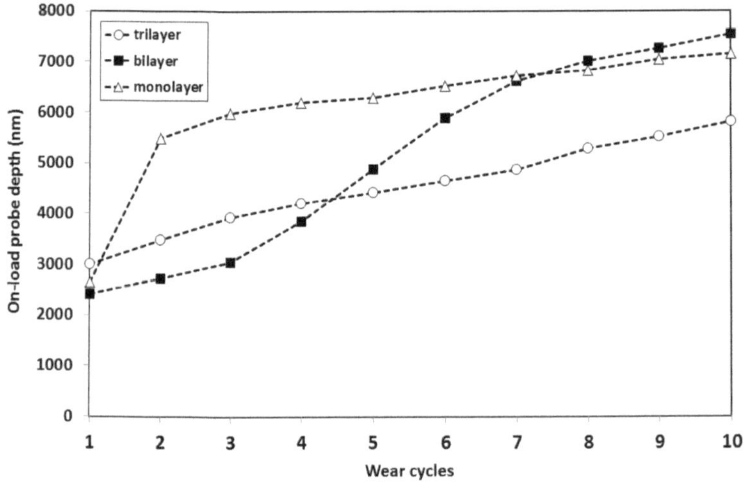

Figure 1.17 Influence of increased thickness and load support on critical load in the microscratch test.

macroscale scratch testing. The critical load for delamination failure was dependent on coating thickness and scratch orientation relative to polishing marks on the surface made prior to coating deposition. High roughness tends to decrease the critical load but the load-carrying capability of higher thickness can have a much greater effect. Despite being much smoother, the critical load on 500 nm TiN was only ~ 50% of the critical load on the 1500 nm TiN. The critical load was 20% lower when scratching perpendicular to the grinding marks than parallel to them.

1.3.8 *Friction*

The friction force in the nanoscratch test (and assuming Amontons' law friction coefficient μ) can be deconvoluted into its interfacial and ploughing components so that the interfacial friction can be reported:

$$\mu_{\text{total}} = \mu_{\text{interfacial}} + \mu_{\text{ploughing}}$$

This can be done by different approaches: (1) performing constant-load friction test at very low force where contact is completely elastic and the ploughing contribution is zero, (2) performing

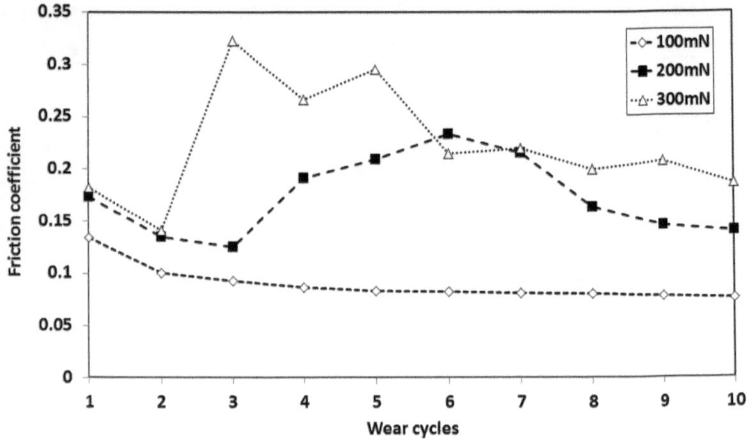

Figure 1.18 Variation in friction with number of wear cycles.

repetitive scratches to eliminate the ploughing contribution, and (3) performing progressive-load scratch and extrapolating the low-load friction data to zero load.

The friction force can be very sensitive to the onset of coating failure, depending on the failure location relative to the moving probe. Figure 1.18 shows the evolution of the friction coefficient with wear cycles at 100–300 mN for 1500 nm TiN on M42 steel being scratched with a 4.4 μm diamond probe. With repeat scratching the friction coefficient initially decreases due to a reduction in the ploughing component. Coating failure was marked by a sudden sharp increase in friction during a single wear cycle (cycle 2 at 300 mN and cycle 3 at 200 mN in Fig. 1.18), which is accompanied by an inflexion in the depth versus wear cycle. After the abrupt increase in friction at film failure the friction decreases with continued scan cycles.

1.4 Nanofretting

Miniaturization of mechanical devices results in severe contact conditions generated by relatively small forces. Performance of materials at small contact scales is a significant challenge and tribology becomes an enabling technology for any small-scale devices with moving components. According to the Amontons' law, the friction force is independent of the apparent area of contact; however, the law does not hold true when the contact area is dramatically reduced. Fretting experiments performed at small contact scales bring specific challenges, including increased importance of surface energy and adhesion, role of wear particles, and impact of surface roughness.

Nanotribology research involving nanofretting investigations tends to focus on two main areas related to (i) fundamental science leading to the understanding of basic principles of friction and asperities contact and (ii) applied research related to technology of small devices like nano- and microelectromechanical systems where the standard constructs of classical physics do not always hold true. So far most of the work conducted on reduced contact scales has been focused on friction; however, with novel measurement and modeling techniques established in recent years, the nanowear problem has been attracting more attention and interesting results are becoming available.

1.4.1 *Fretting Damage*

Fretting is considered as a specific type of reciprocating sliding. It is defined as a small displacement amplitude oscillatory motion (usually induced by vibrations) between two nominally stationary solid bodies in contact under normal load. Depending on the loading conditions (displacement amplitudes, normal loading) fretting causes damage by surface fatigue or wear induced by debris formation. Fretting is indeed a complex process and not all the kinetics and mechanisms are well known. Collins defined more than 50 factors that influence fretting [60] and more factors are still being added [61].

Two major test configurations have been developed for fretting experiments (Fig. 1.19):

(i) Fretting fatigue configuration, derived from a classical fatigue test where two pads are pressed on the fatigue specimen. Relative displacement (δ) and cyclic tangential load (Q) are

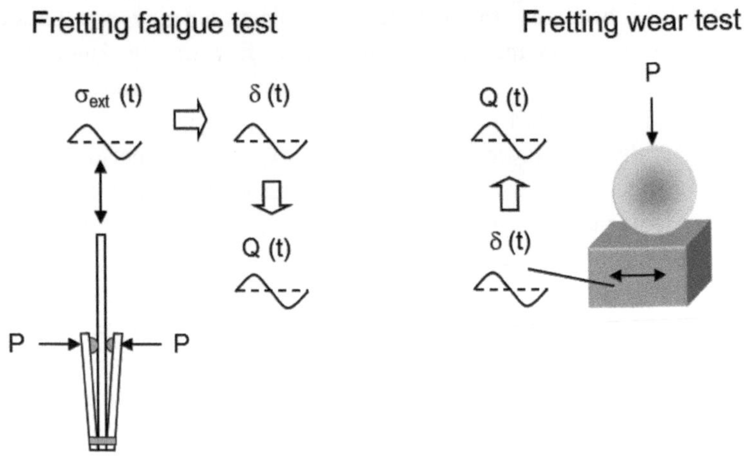

Figure 1.19 Fretting test configurations.

induced by the strain generated through the fatigue part subjected to external loading. In this configuration crack nucleates and then propagates due to applied external loading.

(ii) Fretting wear configuration, where contact load is generated by a relative motion between two bodies induced by an electrodynamic shaker or a hydraulic system. This configuration is usually applied to analyze the wear process due to microdisplacements in a classical reciprocating wear test.

The most established fretting test configuration used in fretting wear experiments is the ball-on-flat one. In such configuration, the ball is subjected to normal load (P) or rotational load (P_r), Fig. 1.20, and three fretting modes can be distinguished [62]: (i) linear displacement (mode I), (ii) radial displacement (mode II), and (iii) circumferential displacement (mode III).

Most published research is focused on fretting mode I, because the wear and friction mechanisms in that case are described by similar laws as sliding friction. However, other modes are also encountered in industrial applications and need to be investigated as well. Fretting mode II occurs in mechanical systems such as ball bearings or electrical contacts, while fretting mode III can

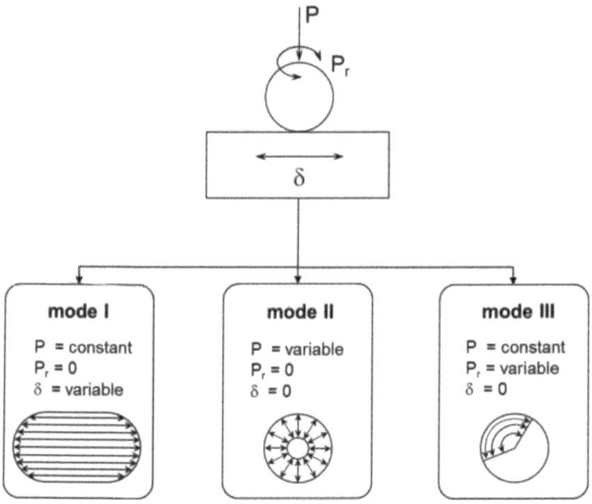

Figure 1.20 Three basic fretting modes with representative displacement tracks.

be observed in the specific areas of heat exchangers or steam generators.

Literature sources quote different values of displacement amplitudes as a border between the fretting process and reciprocating sliding motion. This value is variously interpreted and contained within a wide range of amplitudes, typically between 50 and 300 μm for classical macroscale experiments [63–66]. Although it is possible to limit the relative displacements between particular elements of mechanical systems by means of engineering measures, it is nearly impossible to eliminate fretting entirely, as this kind of degradation was reported even for displacement amplitudes less than 1 μm [67, 68]. The upper limit of the fretting regime can be determined using the coefficient introduced by Fouvry et al. [69], which is defined as a sliding ratio: $e = \delta_g/a$, where δ_g = sliding amplitude, which is different from the displacement amplitude due to the contact and testing device compliance; a = contact radius. The tribosystem remains in the fretting regime when the unexposed surface is maintained at the center of the fretted surface ($e < 1$). The system moves into the reciprocating sliding regime when the center of the contact area becomes exposed to the atmosphere ($e > 1$) (Fig. 1.21).

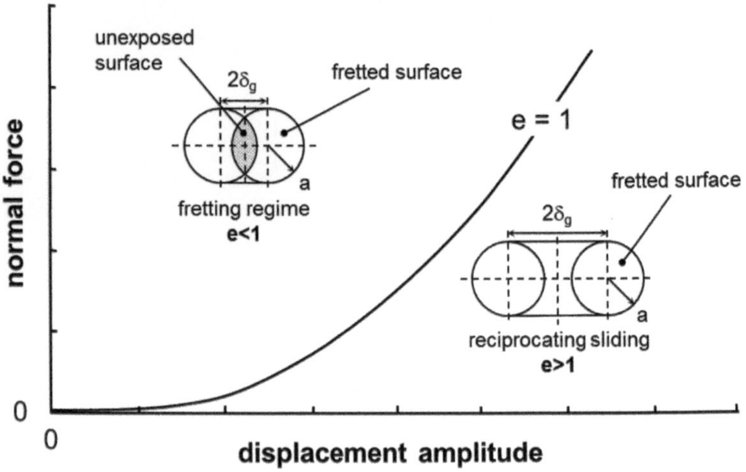

Figure 1.21 Definition of sliding ratio *e* and identification of transition between fretting regime and reciprocating sliding motion (contact sphere/plane). After Ref. [69].

1.4.2 *Small-Scale Fretting*

Analytical tools for nanoscale materials have rapidly developed over the last three decades, improving resolution and efficiency of existing techniques and enabling development of new tools. Starting with the invention of the scanning tunneling microscope in 1981, various experimental methods based on scanning probe microscopy (SPM) have revolutionized many areas of science and engineering by offering a wide range of resolution from the subatomic level up to few hundreds of micrometers. Among various SPM techniques, AFM has contributed to the opening of a new chapter in the field of study of interaction of surfaces in relative motion at small scales, allowing for precise tests to be carried out under ultralow loads and minute sliding distances. Nevertheless, it has been also shown [70] that AFM techniques are not fully suitable for testing MEMS applications as they offer much smaller sliding speeds compared to the real MEMS devices. This is where other novel microscale mechanical testing methods fill the measurement gap, which has been identified between conventional tribometers and AFM techniques (Fig. 1.22) [71].

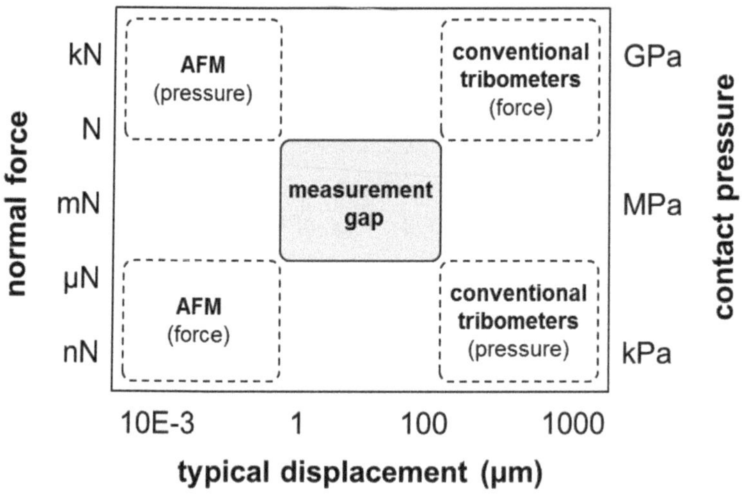

Figure 1.22 Measurement gap between conventional tribometers and AFM techniques as function of normal force, displacement amplitude, and contact pressure. After Ref. [71].

Following the nanotechnology definition, nanotribology is typically defined as a branch of tribology that deals with wear dimensions of less than 100 nm. Currently, there is no formal nanofretting definition available; however, on the basis of the nanotribology description and sliding ratio criterion introduced by Fouvry et al. [10], nanofretting can be defined as "small displacement amplitude oscillatory motion between two nominally stationary solid bodies in contact under normal load, with the unexposed surface maintained at the center of the fretted surface resulting in wear depth of less than 100 nm." This definition is limited to the mode I fretting regime; however, it can be easily adapted to modes II and III by modifying the aspect of the unexposed surface area within the wear track. Although this definition captures the nature of the nanofretting regime clearly, most of the current literature claiming to report nanofretting results would probably not fit into it. For the purpose of simplicity, these types of experiments are described as small-scale fretting in this book chapter, although they could also be described as nano-/microwear, for example.

Small-scale fretting data have been reported by several research groups carrying out experiments on various commercially available and purpose-built equipment. Varenberg et al. studied partial- and gross-slip fretting behavior of 3.1 μm diameter SPM probes tested against Si wafers [72]. A substantial increase of the coefficient of friction was noted at the transition between partial- and gross-slip fretting regimes by testing a range of sliding amplitudes from 5 to 500 nm. In collaboration with the same research group, Yoon et al. investigated the impact of surface roughness and materials' combination on fretting wear in a microspherical contact using a 200 μm diameter spherical tip under a 20 mN load and a 6.5 μm displacement amplitude [73]. It was found that the materials' combination has a greater effect on fretting wear behavior than surface roughness in these small-scale experiments.

Small-scale fretting behavior of monocrystalline silicon for potential application in MEMS devices operating in vacuum conditions was studied by Yu et al. using atomic force microscope tips [74, 75]. The energy ratio related to the transition from the partial- to the gross-slip regime was measured and compared to the same energy ratio observed in classic macroscale fretting. The authors looked also at 2 nm thick DLC coatings deposited on Si(100) and carried out tests against SiO_2 microspheres under vacuum and air conditions [76]. They found that DLC coating reduced significantly adhesion and friction force in air conditions comparing to a Si(100) substrate; however, wear behavior was quite similar to that of Si. Wilson et al. focused on C- and Cr-doped amorphous C films and carried out small-scale fretting experiments using a modified nanoindenter (NanoTest system) with a 300 μm diameter ruby tip under a 10–200 mN applied load and a 2–14 μm displacement amplitude [77–79]. The authors identified two distinct fretting wear regimes, with a classic W-shaped wear scar under low oscillation amplitude and a full U-shaped wear scar at larger amplitudes. Finally, microwear behavior of DLC and TiN coatings using a microtribometer setup under reciprocating sliding was investigated by Achanta et al., where the wear mechanism and topographical changes within the wear tracks were investigated using AFM analysis [80, 81].

1.4.3 *Indentation, Scratch, and Small-Scale Fretting on Si*

There has been considerable interest in the mechanical characterization of silicon for many years; however, highly complex mechanical and tribological behavior of silicon-based MEMS technology remains a main challenge. Nanoindentation of silicon has been the subject of a number of rigorous studies [82–85], but less is understood about its behavior under more complex mechanical situations. Small-scale tribology experiments require high-resolution equipment with ultralow drift, ideally 0.005 nm/s or less, to allow measurements of wear occurring at low contact pressures over periods of one hour or more. Typically AFM methods have been used for such experiments; however, this piezo-based technology cannot provide necessary stability for long duration tests [86]. A number of microtribometers have been adapted for small-scale tribology experiments but these are usually based on cantilevers to measure forces, which do not provide enough stiffness to carry out tests under very low displacement oscillations [87, 88]. Hence, nanoindenter platforms have been recently modified for different types of reciprocating testing under a milli-Newton range constant or ramped applied normal loads [88].

In [55] such a modified nanoindenter platform was used to perform small-scale indentation, scratch, and fretting tests on highly polished Si(100) using a 4.6 μm spheroconical diamond indenter. Figure 1.23 shows that loading curves in all three experiments followed the same trend below a 40 mN applied load. Above that load, tangential loading in scratch and fretting experiments promoted yield, resulting in higher penetration depths. SEM analysis of silicon samples revealed that pronounced lateral cracking observed under scratch and indentation conditions at high loads was absent in fretting tests, suggesting that the accumulation of strain is reduced by the more dynamic wear process. Critical loads observed in the three types of contact are summarized in Table 1.2.

In Ref. [88] small-scale scratch and fretting methods were compared using 10 and 25 μm end radius spheroconical and Berkovich probes on a range of thin-film samples. It was shown that

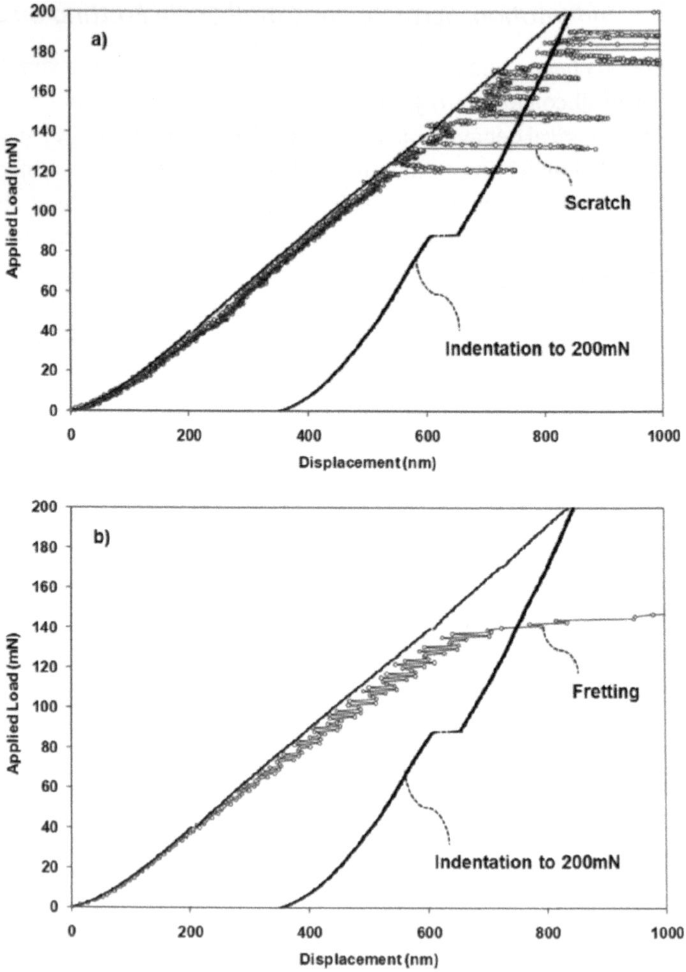

Figure 1.23 Comparison of loading curves in indentation, scratch, and fretting experiments. Test configuration: 4.6 μm spheroconical diamond indenter against Si(100). Adapted from Ref. [55].

small-scale fretting capability is complementary to scratch testing as it offers higher-cycle tests with extended duration and significantly lower contact pressure. Table 1.3 summarizes the key differences between the techniques.

Table 1.2 Critical loads in small-scale indentation, scratch, and fretting experiments

	Critical load (mN)		
	Indentation	Scratch	Fretting
L_y	40 ± 5	37 ± 5	~ 30
L_{c1}	$\sim 80–95$	56 ± 4	*
L_{c2}	>200	106 ± 7	*
L_{c3}	>300	155 ± 16	$\sim 100–110$

*With the faster speed in the fretting test the L_{c1} and L_{c2} transitions are smeared out and cannot be clearly defined.
Source: After Ref. [55] with permission.

Table 1.3 Comparison of conditions and contact pressures in small-scale scratch and fretting of amorphous carbon films on Si

	Ramped scratch	Repetitive scratch	Fretting
Motion	Unidirectional	Unidirectional	Reciprocating
Probe radius	1–9 μm	1–9 μm	10–25 μm
Wear cycles	1	5–20	to 10,000
In situ wear measurement	Yes	Yes	Yes
Contact pressure	8–15 GPa	9–13 GPa	0.1–6 GPa

Source: Adapted from Ref. [88].

1.4.4 *Small-Scale Fretting of ta-C Films on Si*

Wear and adhesion limit the practical usability of silicon-based devices and several approaches are being considered to overcome that problem, including liquid lubrication [89], solid lubrication with self-assembled monolayers [90], and a variety of thin films [91–93]. In Ref. [48] small-scale fretting of 5, 20, and 80 nm thick ta-C films deposited on Si(100) was investigated using spherical indenters. It was found that fretting wear occurred at significantly lower contact pressure than is required for plastic deformation and phase transformation in indentation and scratch experiments. Energy-dispersive **X**-ray spectroscopy (EDX) analysis performed across fretting wear scars revealed a reduction in the coating thickness and oxygen incorporation in the worn region. The absence of any abrupt changes in depth or friction showed that deformation proceeded by a fatigue mechanism with a gradual wearing away of the film. SEM analysis also suggested distinctive gross-slip type of damage with scratches generated during sliding present within the fretting wear scar area. The geometry of the wear scars correlated with the loads applied: reduction in wear track length and wider wear tracks were observed with increased load.

1.4.5 *Small-Scale Fretting of Biomedical Materials*

Artificial joints are designed as fully lubricated systems; however, it has been found that real contact area is only a small fraction of nominal contact area in such devices and asperity-to-asperity contact will occur, leading to metallic contacts at the microscopic level [94]. Surface scratches observed in retrieved prostheses are typically in the range of a few microns wide and less than a micron deep [95], so the mechanical and tribological properties of biomaterials need to be investigated at the relevant contact scale in order to develop reliable artificial joint design. Hence, in Ref. [96] small-scale fretting experiments were carried out on biomedical grade Ti6Al4V, 316L stainless steel, and CoCr alloy samples using

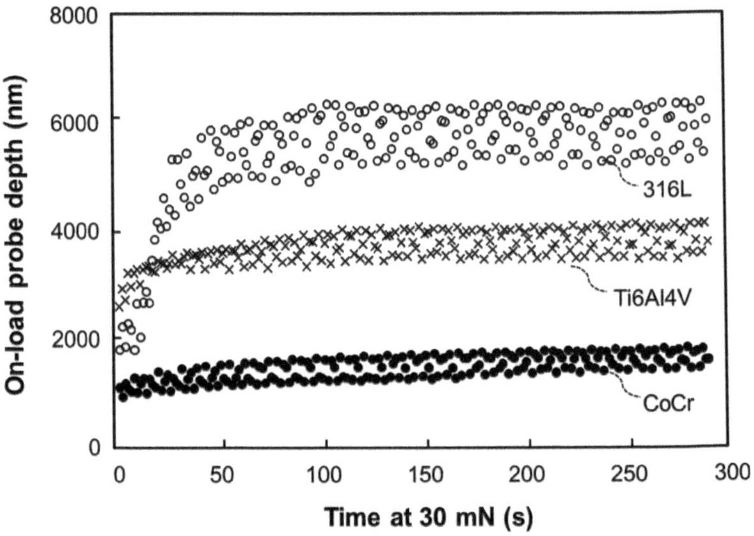

Figure 1.24 On-load probe depth evolution as a function of 30 mN small-scale fretting test. Adapted from Ref. [96].

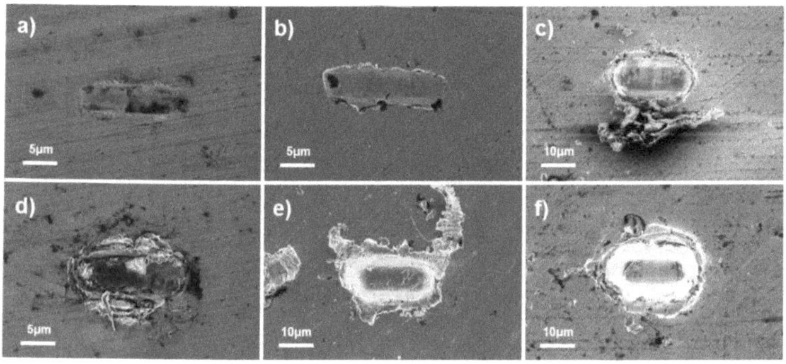

Figure 1.25 SEM images of small-scale fretting scars: (a) CoCr alloy under 3 mN load, (b) Ti6Al4V alloy under 3 mN load, (c) 316L stainless steel under 3 mN load, (d) CoCr alloy under 30 mN load, (e) Ti6Al4V alloy under 30 mN load, and (f) 316L stainless steel under 30 mN load. Adapted from Ref. [96] with permission.

a 3.7 μm spheroconical diamond indenter. Over a wide range of experimental conditions the CoCr alloy exhibited significantly better wear resistance comparing to 316L steel and Ti6Al4V for which wear resistance markedly deteriorated as the fretting load increased. Figure 1.24 shows the probe depth under load in the 30 mN small-scale fretting experiment during the 290 s at peak load. At the beginning of the test 316L steel performed better than the Ti6Al4V alloy, which was followed by an abrupt increase in probe depth on the stainless steel. This type of behavior was observed in other tests for the Ti alloy and 316L steel but not on the CoCr alloy. Figure 1.25 presents the corresponding SEM images, which show that debris on CoCr appeared generally smaller in size, while Ti6Al4V and 316L stainless steel showed delamination type of wear. Small-scale fretting experiments proved to be a valuable additional tool in this work, extending the capability of nanomechanical testing of biomaterials.

1.5 Nanoimpact

In many engineering systems materials are subjected to high-strain-rate deformation. Examples include interrupted cutting (milling), contact between valve-train components in internal combustion engines, and collisions between air-borne particulates and gas turbine blades [26, 27, 52, 97–99]. The ability to simulate the relevant in-service contact conditions is highly desirable in the development of materials for these applications and the nanoimpact technique is increasingly being used for this purpose. The nanoimpact technique works by subjecting the sample material to either a single or a repetitive high-strain-rate contact with a test probe. Experiments can be closely defined, the user having control over a wide range of parameters:

- Probe material and geometry
- Force with which the probe is accelerated toward the sample
- Distance over which the probe is accelerated
- Either a single or a repetitive contact
- Frequency of the contacts and the total experiment time if a repetitive contact is used

1.5.1 *Single-Impact Tests*

Single-impact experiments are typically used to determine the "dynamic hardness" of a sample [100, 101]. This is the effective hardness of the material at a given rate of strain. The dynamic hardness is determined by the quantity of energy absorbed into the deformed material volume at the contact site. The required data for this calculation are provided by a penetration depth versus time trace, which is recorded over a very short time period (typically <0.2 s) as the impact takes place. From these data the incoming and outgoing velocity of the indenter are determined as is the maximum depth reached. A dynamic hardness value will almost always be higher than the quasi-static hardness value for any given material. This is due to the extremely short contact duration between the probe and the sample, meaning that minimal time-dependent deformation can take place (Fig. 1.26).

The above data were gathered from a single impact experiment on a Ti6Al4V alloy. The dynamic hardness in this test was 8 GPa compared to around 5 GPa under quasi-static conditions. Constantinides et al. [101] reported that gold was even more

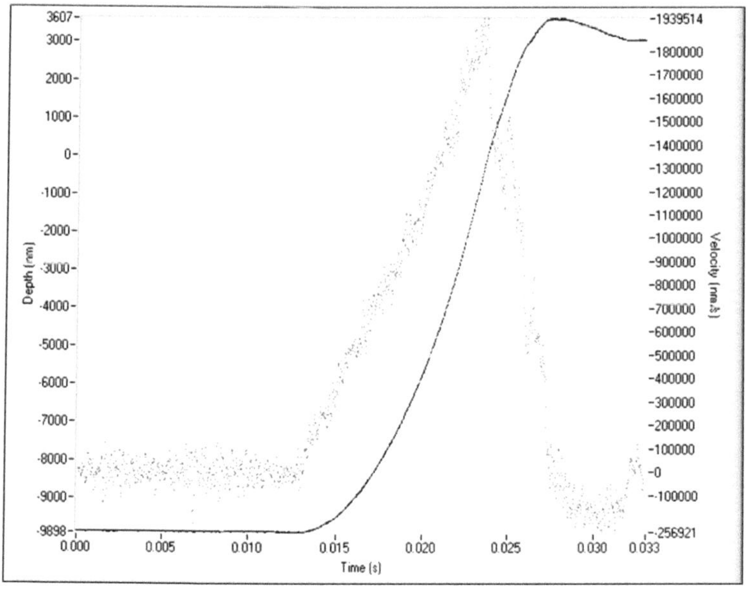

Figure 1.26 Time–displacement and time–velocity data from a single-impact event on Ti6Al4V. The data show the acceleration of the indenter, the contact with the sample, the deceleration of the indenter, and the first rebound of the indenter.

sensitive to the strain rate but aluminum was much less so (Table 1.4).

Table 1.4 Comparison of dynamic and static hardness of metals

Sample	Quasi-static *H* (GPa)	Dynamic *H* (GPa)
1100 aluminum	0.27	0.29
annealed Au	0.48	1.3
work hardened Au	0.80	2.8

Source: After Ref. [101].

1.5.2 *Repetitive-Impact Tests*

Repetitive-impact tests are typically used to assess the ability of a sample to resist failure due to fatigue. The technique is essentially comparative and allows samples to be ranked in terms of their performance under the test conditions. As in the single-impact test, the resulting data take the form of a penetration depth versus time trace, but a typical time frame may be 60–300 seconds. Cross referencing with an optical or atomic force microscope the depth versus time data of the impact site can reveal a great deal about the cause and mode of failure(s). A popular application of the repetitive-nanoimpact test is to test coating properties at high strain rates and to investigate surface fatigue and fracture due to repetitive contact, often with the aim of using results to design and optimize the coating properties for improved durability in interrupted contact situations [26, 27]. In practice there is usually a direct correlation between increased fracture resistance in the nanoimpact test and longer lifetime in the application.

This correlation can also extend to the temporal evolution of impact wear. For example, Fig. 1.27 reveals the strong similarity between the evolution of tool wear in cutting of 1040 structural steel and the progression of coating damage on repetitive-nanoimpact testing. With its lower PI the TiAlN fractures more extensively in the nanoimpact test. Plotting data as fracture probabilities allows for the stochastic nature of the impact fracture process to be investigated [102].

Typically the depth evolution is gradual, with periodic abrupt increases in probe depth being due to fracture. Using the NanoTest system modified for fast data acquisition Jennett and Nunn showed that there were clear differences in the energy absorbed when fracture occurs [103]. Figure 1.28 shows repetitive-impact data on fused silica at 8 mN with a cube corner diamond indenter. It clearly shows that the impact resulting in a large abrupt change in depth is correlated with enhanced energy absorption. Recently the nanoimpact behavior of a Ti0.4Al0.6N coating was simulated with a 3D finite element model (FEM) by Bouzakis et al. [104]. The model used the failure of constrained tied nodes to simulate crack formation and propagation as the plastic strain develops and exceeds the coating failure strain. The developed nanoimpact FEM simulation was able to predict the film failure initiation and evolution with repetitive impact with a cube corner indenter at 10–50 mN impact force. With spherical probes at low impact forces so

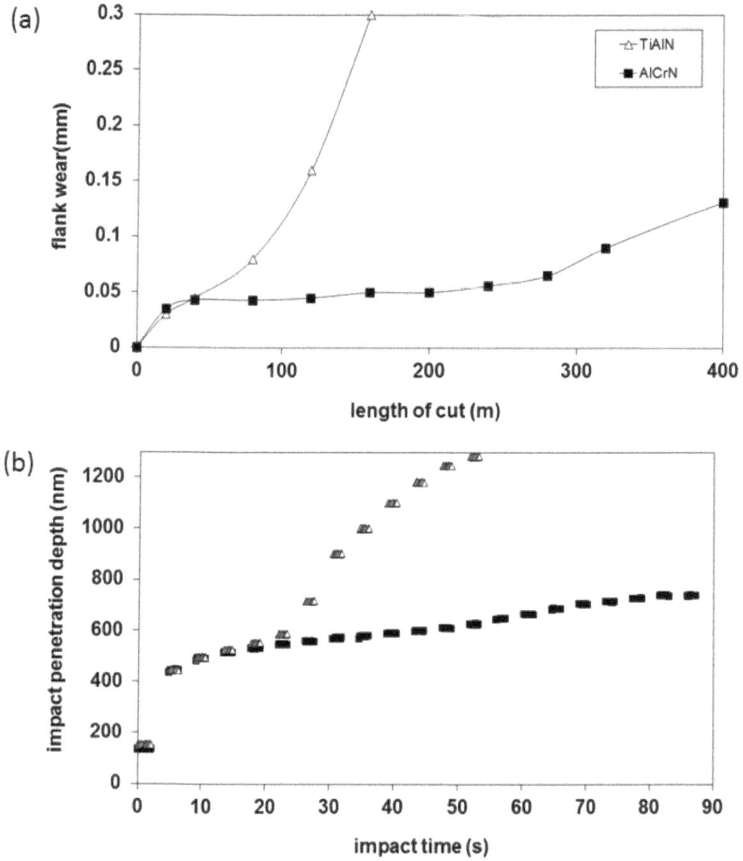

Figure 1.27 Comparison of the performance of TiAlN and AlCrN coatings in (a) a cutting test and (b) a nanoimpact test.

that plasticity is suppressed, a backward depth evolution can be observed (i.e., the on-load probe depth decreases with continued impacts) [105, 106]. This atypical behavior has been confirmed by post-test AFM imaging of impact scars showing volume uplift and has been interpreted as being due to interfacial delamination without (or before) the accompanying fracture that results in the increase in depth. Faisal et al. reported that in the repetitive impact of a 100 nm DLC film on Si the geometry of the impact probe

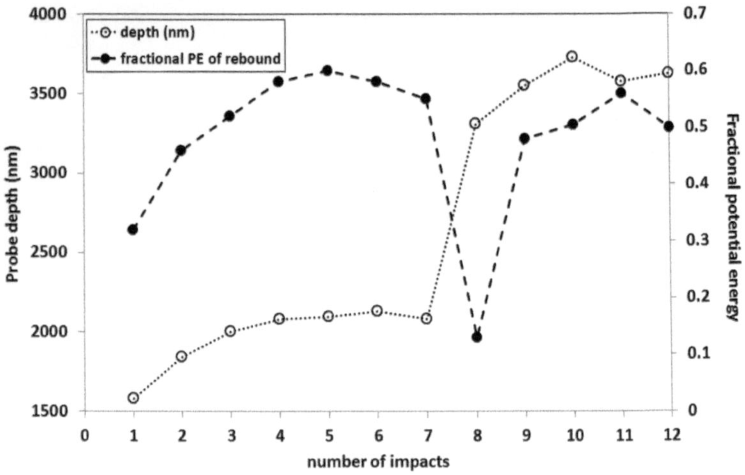

Figure 1.28 Correlation between fractional potential energy and probe depth in the repetitive-impact test. After Ref. [103].

influenced the deformation with a conical indenter resulting in delamination but Berkovich resulting in fracture [105, 106].

The following example shows the relative-impact resistance of two TiO_2-coated stainless steel samples. Both stainless steel substrates have been sand-blasted, and one of them was subsequently given a heat treatment process before deposition of the TiO_2. Each test was 5 min in duration and during this time 75 repeat contacts were made at 0.25 Hz, the indenter being allowed to contact the sample for 2 s then removed from the sample for 2 s.

The effect of the heat treatment of the substrate is clearly evident in the depth versus time data and microscopy. Depth "jumps" and regions of exposed substrate are visible in Fig. 1.29a,b. The heat treatment was shown to improve the resistance to failure. The smaller maximum depth attained suggests that the improvement was likely to be due to an increased hardness of the substrate helping to provide a better load support for the coating.

The nanoimpact test has been used to assess the influence of the deposition of thin interlayers of Ni or Cr (15–50 nm) before depositing 250 nm Pt–Ir coatings on binder-less cemented carbide tools used for precision glass molding [52]. Using a cube corner

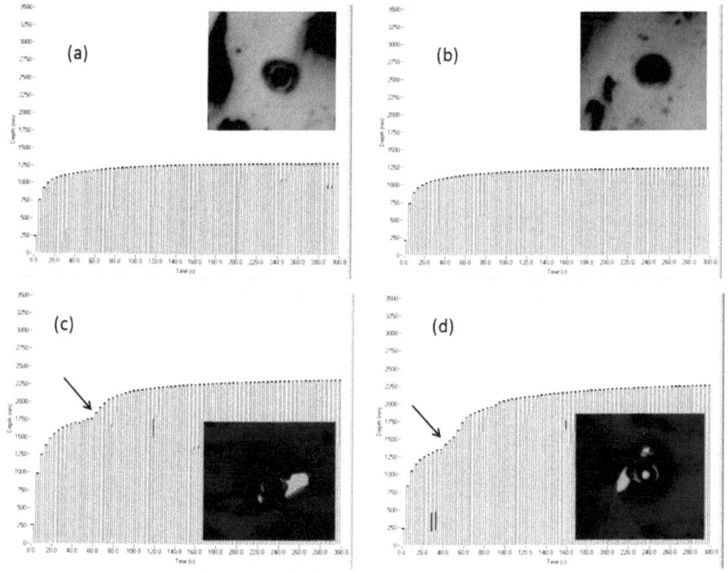

Figure 1.29 Illustrative repetitive-nanoimpact depth vs. time traces and optical microscopy of TiO_2 coating on (a and b) heat-treated sand-blasted stainless steel substrate and (c and d) sand-blasted stainless steel substrate without prior heat treatment.

probe it was found that on coatings without interlayers the probe depth continued to increase dramatically as the repetitive-impact test progressed but coatings with interlayers were significantly more impact resistant. The best impact resistance was shown by a Pt–Ir coating with a 50 nm Cr interlayer where after initial plastic deformation the probe depth remained approximately constant to the end of the test.

Bouzakis et al. have performed detailed investigations into the influence of postdeposition wet microblasting on cutting performance of PVD-coated tools, supporting their tool life data with nanomechanical testing and FEM [107–109]. Microblasting induces compressive stresses, which can be beneficial in reducing wear, although the results are dependent on pressure and the grain size and geometry of the abrasive materials. For the same microblasting conditions and abrasive diameter the abrasion with ZrO_2 is less intense than Al_2O_3 due to the spherical nature of ZrO_2 [107]. To

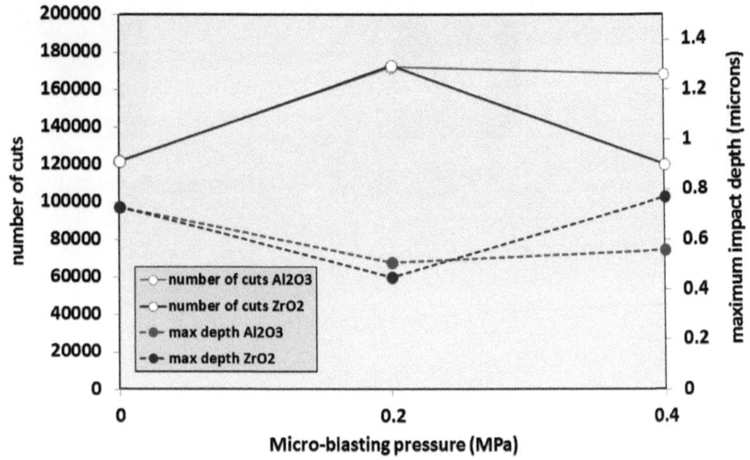

Figure 1.30 Lower impact depth correlates with greater cutting performance.

investigate the influence of the developed compressive stresses on the film brittleness the researchers performed nanoimpact tests, finding impressive correlation between the results of the nanoimpact tests and the cutting performance when milling hardened steel (AISI 4140) [108].

Figure 1.30 shows clearly that under these conditions (i) low impact depth (high impact resistance) is strongly correlated with cutting tool life, (ii) the trends in tool life with microblasting pressure are well reproduced by the nanoimpact test, and (iii) the ranking of microblasting with ZrO_2 and Al_2O_3 at a given pressure and the switch in relative performance between 0.2–0.4 MPa are also shown.

Chen et al. highlighted the importance of microstructure on the dynamic impact response in nanoimpact of wear-resistant hard TiAlSiN and TiN thin coatings on M2 steel [110]. When the impact load was increased to 10 mN and 30 mN for the monolayer TiN and multilayer TiAlSiN coatings, respectively, cracks started to appear during the tests. At higher loads, spallation was observed in both coatings, although the multilayer TiAlSiN coating showed much less spallation. The improved nanoimpact resistance of TiAlSiN coatings is considered to be due to its H^3/E_r^2 aiding resistance to crack formation and its multilayer structure helping with crack deflection.

Chen et al. investigated the role of changing coating microstructure and mechanical properties due to thermal aging at 1100°C–1500°C on the repetitive-impact and erosion behavior of 150 μm thick thermal barrier coatings (TBCs) for aeroengines [99]. Thermal aging alters the microstructure and mechanical properties of TBC systems, and there is a clear need to develop a rapid laboratory test method that can assess such changes. Nanoimpacting provides such a dynamic measurement tool and in this study it has been used to assess erosion resistance. For this application it is superior to conventional nanoindentation as the dynamic strain rates are much higher. Compared to bulk erosion testing it offers the potential advantage of testing on the scale of the individual columns in the TBC, thus measuring individual column and column cluster properties. It is shown that thermal aging resulted in increased hardness and stiffness of the TBC. Nanoimpact testing (Fig. 1.31) showed the increased wear rate very clearly under these conditions and this correlated well with the measured bulk erosion rate. As a repetitive dynamic test producing damage on the correct length

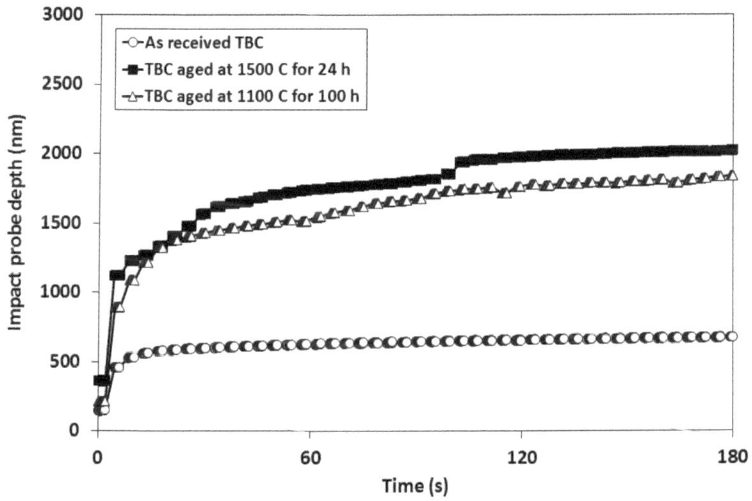

Figure 1.31 Influence of thermal aging for 24 h at 1500°C and 100 h at 1100°C on the impact resistance of an EB-PVD TBC.

scale, the nanoimpact test was highly effective in simulating the erosion mechanism and may be used as a rapid screening test to evaluate the erosion resistance. Nanoindentation alone would not be able to predict this.

1.6 Environmental Considerations (Hot, Cold, Fluid, Humidity)

Nanoindenters most commonly measure sample mechanical properties within a fairly narrow window of temperature (e.g., 20°C–25°C), humidity (relative humidity [RH] $\approx$ 40%–50%), and strain rate ($\sim 10^{-1} - 10^{-3}$ s^{-1}). However, there are many situations in which characterization under these conditions is simply not appropriate and it is more important to measure properties under conditions more reflective of the operating or processing conditions that the test material is to be exposed to. Mechanical and tribological behavior under these conditions may bear very little relation to that measured at room-temperature, quasi-static conditions, and attempting to use data to predict performance under very different conditions can be highly dangerous. Originally nanoindenters could only operate within this confined window of humidity, strain rate, and temperature; however, developments over the last 10 years or so have seen rapid progress in expanding the test capability of some commercial instruments. Environmental extremes (high temperature or in liquid) typically act to promote time and temperature dependence, which alters the indentation response, taking it either further from or closer to elastic unloading. Careful experimental design is required.

1.6.1 *Influence of Moisture and Electrochemical Potential on Nanoindentation Behavior*

Performing measurements in controlled humidity conditions requires either placing the entire instrument in an environmental chamber (or modifying its standard operating test enclosure) or designing a small humidity control unit to enclose a much smaller volume immediately in the vicinity of the sample. An example of the former approach was by Altaf et al., who recently used the NanoTest system with its test enclosure fitted with a humidity control unit to study the effect of moisture on a commercial stereolithography polymer resin, Accura 60, over a wide humidity range (33.5%–84.5% RH) [111]. Mechanical properties of stereolithography resins such as Accura 60 are affected by the level of moisture in the environment, with hardness and modulus decreasing with increased moisture uptake in the resin. Transport of moisture from the surface to the bulk of the polymer took place sufficiently slowly that a coupled stress–diffusion FEA was required to accurately model the behavior. Gravimetric tests were performed to calculate the diffusion constants and bulk tensile, compressive, and creep tensile tests to generate the mechanical material properties for the model. With appropriate modeling, the variation in hardness with (i) increasing penetration into the interior of the polymer and (ii) different environmental conditions could be accurately simulated.

Measurements under fully hydrated conditions require both sample and indenter be fully immersed in a dedicated liquid cell [112–114]. A consideration when measuring a very highly compliant sample in a fluid environment is the buoyancy of the indenter and measurement head components. Any change in buoyancy throughout the measurement, although small, can be significant when the sample material is so easily deformed. In the NanoTest system the indentation direction is horizontal; as the indenter penetrates the sample surface the submerged volume of the loading head remains constant (due to it moving horizontally in the fluid) and as such constant buoyancy is maintained.

Nylon-6 samples with ~30%–35 % crystallinity were tested by Bell et al. in fully hydrated (deionized water) and ambient laboratory (50% RH) conditions using the cell design in Fig. 1.32 [113]. After equilibration (>24 h) the hardness and elastic modulus measured in deionized water were significantly lower (50% and 60%–70% respectively) than those measured in ambient conditions. The strain rate sensitivity parameter, $A/d(0)$, where A is a constant determined by fitting to experimental data and $d(0)$ is the depth at the start of the hold period at peak load, decreased significantly in water. The reduced proportion of time-dependent deformation in deionized water is consistent with a reduction in the tan delta

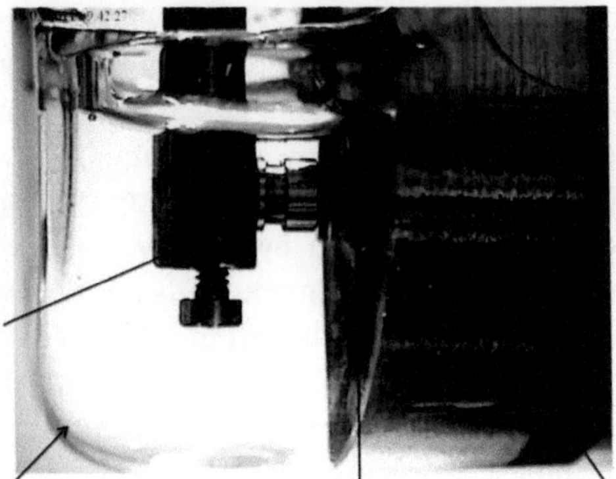

Figure 1.32 The configuration of a commercial liquid cell (NanoTest system).

peak due to a shift in the glass transition temperature when wet. Using the NanoTest system fitted with a recently designed humidity cell, Harris et al. [115] were able to more rapidly measure the moisture sensitivity of the hardness and stiffness of a similar nylon sample over the humidity range 20%–95% RH. Figure 1.33 shows the variation in elastic modulus with humidity.

Making measurements in fully hydrated conditions is particularly useful in the characterization of biological materials. Constantinides et al. [114] found the stiffness of polyacrylamide (PAAm)-based electrophoresis gels decreased by a factor of about 1000 when hydrated: E (gel, water) = 270 kPa; E (gel, air) = 300 MPa. Even larger changes have been observed for polyethyleneimine hydrogel. When indented in its dry state this polyethyleneimine hydrogel exhibits a reduced modulus of almost 18 GPa; after 30 min water immersion this has fallen by a factor of nearly 5000 to 3.3 MPa (Fig. 1.34).

Schmidt et al. used the NanoTest system modified to act as a electrochemical cell to study an electroactive polymer nanocomposite thin film containing cationic linear polyethyleneimine and 68 vol.% Prussian blue nanoparticles as a candidate stimulus-responsive polymer material [116]. Electrochemical reduction of the Prussian blue particles doubled their negative charge causing an influx of water into the film to maintain electroneutrality. This resulted in

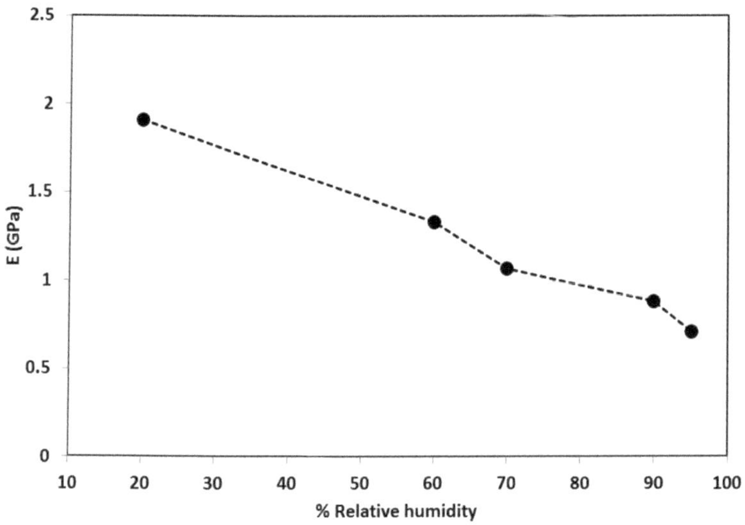

Figure 1.33 Influence of humidity on the elastic modulus of nylon.

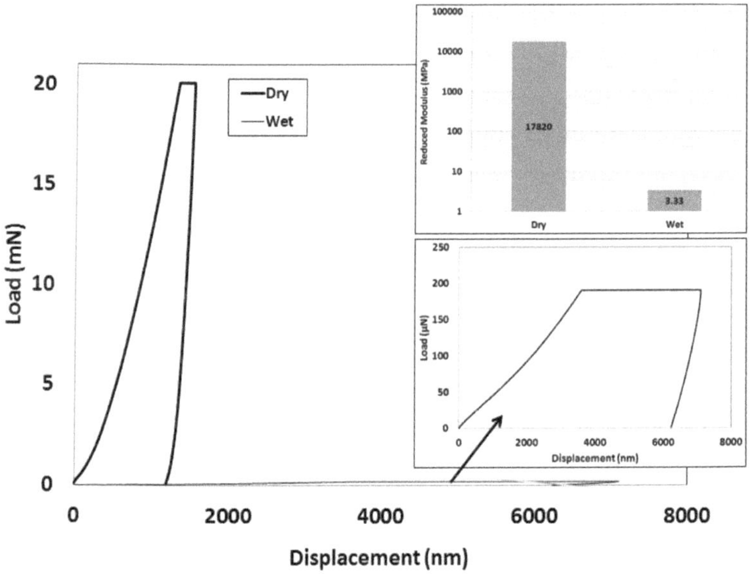

Figure 1.34 Indentation behavior of a polyethyleneimine hydrogel in fully hydrated and ambient air.

swelling and a decrease in elastic modulus. The in situ nanoindentation measurements using a spherical ruby indenter of 5 μm radius showed a reversible decrease in the elastic modulus of the film from 3.4 GPa to 1.75 GPa.

1.6.2 *Nanoscratch and Nanoimpact Testing in Liquid*

The ability to perform nanoscratch tests under liquid has many biological and biomedical applications. Running nanoscale tribological tests under precisely controlled conditions that in terms of stresses, contact size, and test environment are close to those in vivo can provide valuable insight into the underlying deformation mechanisms. As an example, abrasion on retrieved CoCrMo hip joints has been reported to be due to third-body wear by detached hard particles in vivo. The presence of small scratches around 10–20 μm wide suggests contact with sharp asperities. To understand the tribology of the scratch formation it is necessary to perform tests at the relevant scale rather than rely on bulk testing. Sun et al. have used nanoscratch testing of cast CoCrMo to simulate single-asperity contact with a small hard carbide particle [117]. Using the NanoTest system fitted with a liquid cell they performed tests at 37°C in 0.9% NaCl using a 10 μm diamond scratch probe at 200 mN. They found that even single scratches could produce significant microstructural changes in both the Co-rich matrix and the hard carbides. They were able to reproduce various features such as slip lines, matrix fracturing, and the formation of nanometric and rhombic grains on the periphery of the scratch that have been found on retrieved MoM cups. Kalcioglu et al. at the MIT used nanoimpacting in liquid to evaluate the high-strain-rate behavior of fully hydrated tissues from liver and heart and candidate tissue surrogate materials, quantifying their resistance to penetration under energy densities relevant for tissue surrogate applications [118]. The energy strain densities in the tests were 0.4–20 kJ/m^3, which is broadly comparable with macroscale impact tests designed to replicate ballistic conditions (15–60 kJ/m^3). They reported that the energy dissipation capacities of fully hydrated soft tissues were well matched by a 50/50 triblock/diblock gel composition.

1.6.3 *Measuring High-Temperature Mechanical Properties by Nanoindentation*

The ability to measure mechanical properties of materials at in-service temperatures has great value. High-speed cutting tools experience temperatures of >500°C in operation due to frictional heating and TBCs for turbine blades can be required to operate at 1000°C or higher. The high-temperature properties of the materials used in such applications cannot necessarily be predicted from room-temperature testing. Elevated-temperature nanoindentation can provide the engineer with invaluable information to help to optimize material properties for their applications. Much progress has been made in high-temperature nanoindentation over the past 10 years. Understandably the very small depths typically attained in nanoindentation mean that the technique can be susceptible to "thermal drift" as sample and/or instrument components change slightly in temperature and consequently expand or contract during the test.

Reliable high-temperature nanoindentation requires that (i) tests be performed without thermal drift or other instrumental factors influencing the raw data, (ii) the elastic contact mechanics at room temperature be successfully applied to situations in which very significant time-dependent nonelastic deformation may occur, and (iii) oxidative wear of tip and/or sample not occur (or if it does, is sufficiently small not to complicate interpretation of the results). Despite some potential drawbacks, as described earlier, fused silica is a popular choice as reference material for high-temperature nanoindentation since it is mechanically homogeneous, shows relatively little creep at elevated temperature, is mechanically homogeneous with elastic properties well characterized by other techniques such as resonance ultrasound, and has quite low thermal conductivity. Practice has shown that it is not usually sufficient to only heat the sample before indentation, as this produces a large thermal gradient between the hot sample and the cold indenter, which greatly influences the measured data, particularly on more conductive samples. When a nonheated indenter is used to indent a heated sample heat will conduct onto the indenter from the sample, causing indenter expansion. Consequently the instrument measures a combination of the indenter penetrating the sample and the dimensional change of the indenter, thus compromising the experiment. The deficiencies of the sample-only heating approach were recently highlighted by Everitt et al. on fused silica and gold [119], Tang and Zhang on Si [120], and Zarnetta et al. on a TiNiCu shape memory alloy [121]. These studies reported elevated-temperature indentation curves dominated by thermal drift when sample-only heating was used. In the latter publication the authors showed that even loading to peak load at 80°C in five seconds was insufficient on the shape memory alloy and suggested instead loading in one second. Even at moderately elevated temperatures very rapid loading appears the best approach to at least partially counteract the thermal gradient from the sample and indenter being at differing temperatures [122–124]. Franke et al. have studied the temperature dependence of the ISE on copper to 200°C with a sample-only heating [124]. Drift rates increased by a factor of $\times$ 70 as the temperature was increased, reaching 7 nm/s at 200°C, so the tests were performed very rapidly in an attempt to mitigate against this and no modulus information was reported. Nevertheless, they were able to show that the ISE on copper was reduced at higher temperatures, implying thermal activation has a major influence on the length scale for plasticity.

The practical solution to this problem is to separately heat both the indenter and the sample to the same temperature to minimize/eliminate the thermal gradient to achieve isothermal contact. The first report of dual active heating on a commercial nanoindentation system (NanoTest system) was by Smith and Zheng in 2000 [125]. Subsequently, this approach has been adopted in around ~60% of high-temperature nanoindentation publications, including virtually all of those at over 200°C. The configuration for this is shown schematically in Fig. 1.35.

Precise temperature control is important. At high temperature even a small thermal differential can compromise the reliability of the experiment. In the example above the indenter was about 5°C cooler than the sample surface (sample temperature 500°C). Heat flow from the sample onto the indenter has caused expansion of the indenter material during the indentation process. The expansion was sufficiently rapid so that during the 30 s hold at maximum

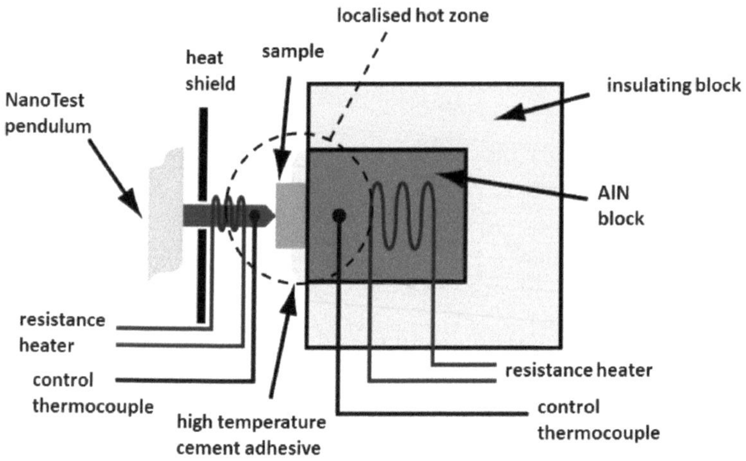

Figure 1.35 Configuration for dual active indenter and sample heating.

load we see a reduction in depth rather than the increase we would expect to observe as the material creeps. Unsurprisingly increasing the indenter/sample temperature mismatch increases the drift dramatically (Fig. 1.36). With a difference of about 20°C initially the depth is seen to increase as the load is applied then, a critical contact area is reached where sufficient heat can flow such that the linear expansion of the indenter is faster than the rate of penetration of the indenter in to the sample material, and consequently the measured depth is seen to reduce.

The effectiveness of the method to minimize (or even eliminate) this thermal gradient by heating the indenter and the sample separately to achieve isothermal contact is improved by precision control. The methodology involves three steps for the indenter heating control side [126]:

(i) The indenter is heated to the target temperature by feedback control.
(ii) Once the target temperature is reached the controller measures the average power supplied to maintain the target temperature.
(iii) The indenter is then supplied with this constant power during indentation.

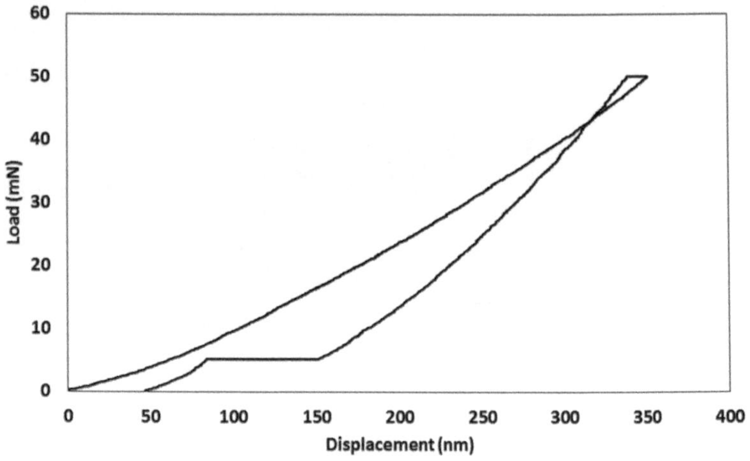

Figure 1.36 The effect of thermal mismatch between indenter and sample due to poor experimental design.

This approach provides more stable temperature control than using PID control for both sample and indenter. Once a suitable indenter has been selected (see below) and the indenter/sample temperatures matched by the above method it is possible to produce measurements of equivalent quality as at room temperature, as illustrated in the following examples. Unless one is familiar with the indentation characteristics of a given sample, it can be difficult to distinguish between high-temperature and room-temperature load–depth curves. The following example shows indentation data on Si (100) at 500°C and 650°C (Fig. 1.37). This is a critical temperature range for silicon over which appreciable softening occurs.

The depths achieved at equivalent loads are markedly different as is the degree of displacement that occurs during the maximum load dwell periods. Materials we consider to be essentially elastic at room temperature often exhibit increasing viscoplasticity with increasing temperature. This can be highlighted by either applying the indenter load at different rates or examining the sample deformation under constant load.

In this example silicon has been indented under two loading conditions, using a constant loading rate (Fig. 1.38a) and a constant loading time (Fig. 1.38b). In case (b), as the loading time for

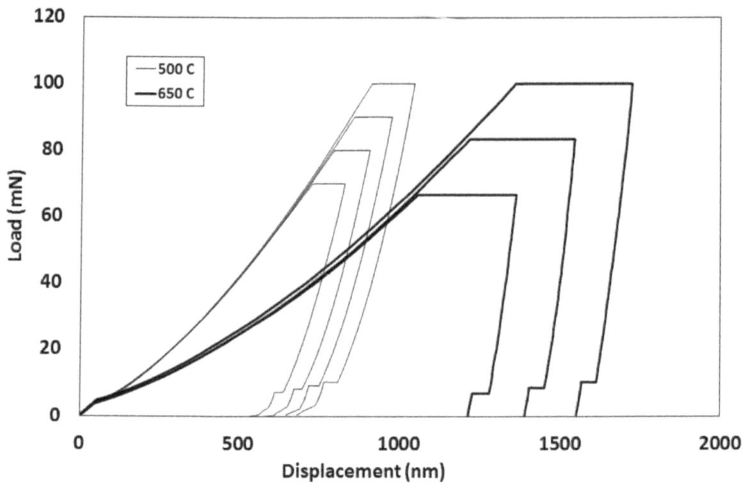

Figure 1.37 Indentations on silicon at 500°C and 650°C.

each indentation was constant the loading rate used increases as a function of the target maximum load, that is, the highest load (200 mN) indentation was loaded at 40 times the loading rate of the lowest (5 mN) indentation. The effect of this is that the loading curves do not superimpose as the time dependent element of the sample's deformation plays an incrementally smaller role in the overall deformation measured. Contrast this with case (a) where the loading curves are seen to superimpose.

1.6.3.1 Tribochemical interactions between indenter and sample

Indentation at elevated temperatures can put a greater demand on the indenter material and consequently care must be taken to select the right indenter material for the sample to be measured if premature blunting of the tip is to be avoided. Blunting of the tip can be caused by (i) oxidation of the indenter material, (ii) dissolution of the indenter material into the sample, and (iii) accelerated wear at temperature due to indenter softening. Oxidative degradation can be eliminated by running indentation experiments in a vacuum environment. Suitable apparatus does exist in a number of cases

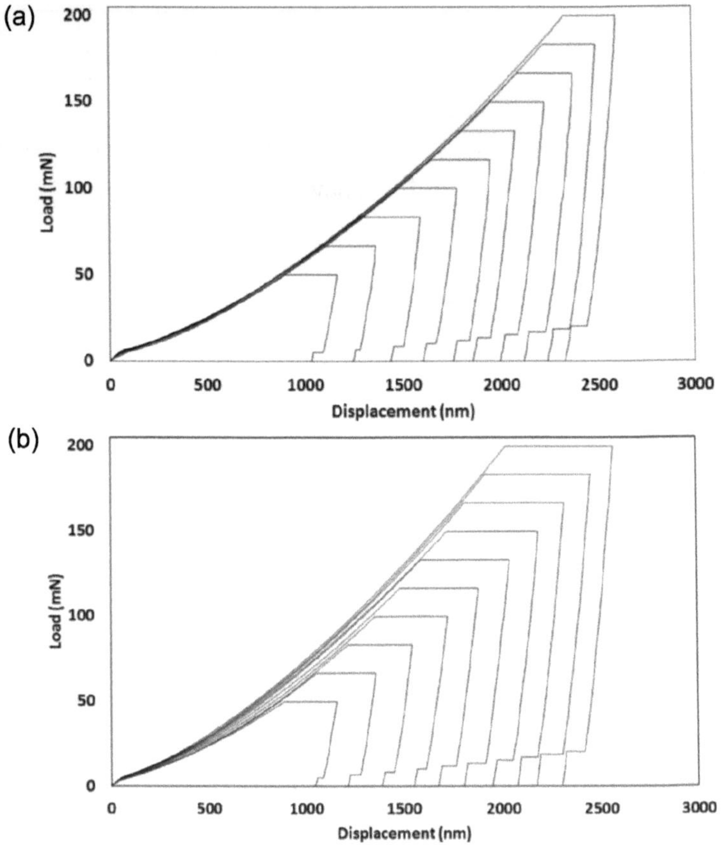

Figure 1.38 (a) Constant loading rate indentation on silicon at 650°C. (b) Constant loading time indentation on silicon at 650°C.

but in addition to cost the integration of nanoindenters into vacuum environments can be challenging as the instruments are susceptible to vibration from pumps and many electronic components found on such instruments are not vacuum compatible. However, where successful integrations exist they can sustain the indenter and the sample at extreme temperatures (up to about 950°C) for extended periods of time with little or no degradation. Purging the instrument environment with and inert gas such as argon can be effective at maintaining a sample in a measurable state. An example of where this can be effective is when cobalt-bonded tungsten carbide samples (coated or bulk) are to be tested at temperatures over 500°C. In air rapid oxidation of the cobalt binder can occur to such a degree that the sample can expand to several times its starting volume. Whilst a purge does not eliminate completely oxidation of such samples it does provide sufficient working time to gather indentation data at temperatures above 700°C.

Carbon is soluble in iron and ferrous alloys. The rate at which it will dissolve increases dramatically with increasing temperature. For this reason a diamond indenter should never be used with such samples as it can be completely degraded in a very short period of time. Sapphire and cubic boron nitride (cBN) are suitable alternative materials for ferrous alloys at elevated temperatures. However, it should be noted that both of these materials have their own limitations. Sapphire is, relative to diamond ($H \approx 100$ GPa), quite soft (about 30 GPa) and indeed in many cases can actually be softer than the sample material—cutting tool coatings for high-speed machining applications can often be over 30 GPa hardness at room temperature. Furthermore, sapphire does not retain its hardness sufficiently to be a viable alternative for measurement of hard coatings at elevated temperatures. It does remain a good option, though, for softer materials such as metals and TBCs. cBN is much closer to diamond hardness (about 70 GPa). It also retains this hardness well at high temperature. The limitations with this material stem mainly from the form in which it is available and any strength-limiting defects that may exist within its microstructure. Typical single-crystal cBN sizes are less than 0.5 mm in any dimension. This presents difficulty in manufacturing for the perspectives of securing the crystal in its holder and achieving the preferred orientation. Despite the difficulties this material presents it is possible to produce thermally and chemically stable indenters that are able to undertake measurements not possible with either diamond or sapphire.

1.6.3.2 (Ti,Al)N coatings for cutting tools: Role of annealing at 700°C–900°C

Coatings have been successfully applied to cutting tool inserts to prolong tool life in machining applications. Trends currently driving the development of new coatings include (i) cutting faster to increase productivity, (ii) "green machining" (cutting without coolant for environmental reasons), and (iii) machining hard-to-cut materials. To achieve a long tool life under these conditions coatings need to be multifunctional and display several interlinked characteristics to minimise wear. For applications such as these involving high temperature due to frictional heating the mechanical properties at the operating temperature are more relevant than those measured at room temperature.

The example below illustrates how annealing coatings can alter their mechanical properties and cutting life. The postdeposition annealing of an $Al_{0.67}Ti_{0.33}N$ (AlTiN) coating, deposited at approximately 600°C, for two hours in vacuum at 700°C–900°C has a dramatic effect on its high-temperature mechanical properties [127, 128]. Figure 1.39 shows the variation in high-temperature hardness with annealing temperature. Without prior annealing the hardness of AlTiN increases slightly at 250°C before decreasing at 500°C. Bouzakis et al. also reported a complex dependence of

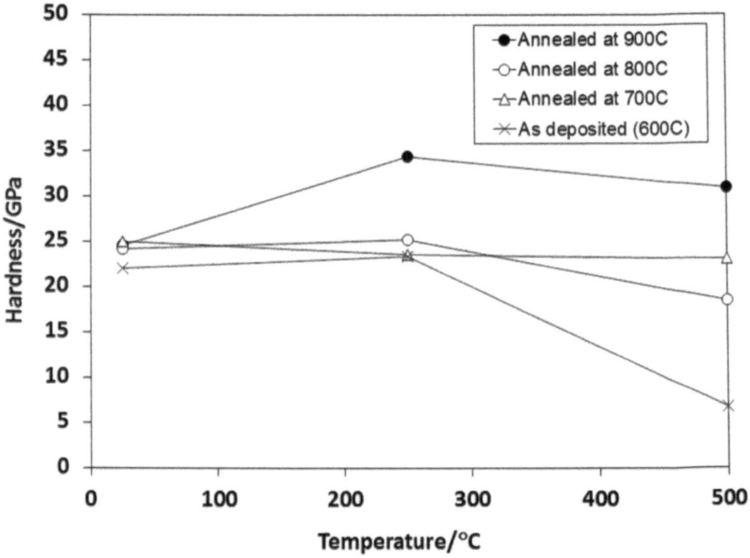

Figure 1.39 Variation in hardness with temperature of as-deposited AlTiN and AlTiN after annealing for 2 h in vacuum at 700°C–900°C.

hardness versus temperature for a (Ti,Al)N coating with similar composition ($Ti_{0.34}Al_{0.66}N$), with a pronounced increase at 150°C–200°C, interpreted as dislocation pileup due to strain hardening [129]. Using dual heating in the NanoTest system [129] they recently reported on the high-temperature nanoindentation of a 3 μm TiAlN coating on WC-Co with high (0.5 μm) surface roughness. The standard deviation and coefficient of variation versus temperature were determined to assess the reliability of the data. The scatter in the indentation depth at 15 mN was unsurprisingly higher than if the sample had been mirror-polished but it increased only minimally as the temperature was increased from 25°C to 400°C.

The influence of the changes of the AlTiN coating mechanical properties from annealing (shown in Fig. 1.39) on the relative life of coated cutting tools in different high-speed machining applications has been studied. The machining operations were continuous high-speed turning of 1040 structural steel, interrupted turning of 4340 steel, and end milling of H13 hardened steel. Annealing at 700°C was found to improve tool life in all these conditions compared to the as-deposited coating. However, when the annealing temperature was increased to 900°C its effectiveness depended strongly on the cutting conditions. The longer tool life in high-speed turning of 1040 structural steel after annealing AlTiN at 900°C has been attributed to the combination of high temperature at the tool/workpiece interface and relatively low stresses at the cutting zone. Under these conditions temperature-induced processes dominate on the surface of the tools and tool life is predominantly controlled by its hot hardness. In contrast, in interrupted cutting conditions toughness and plasticity appear more critical. In end milling high hot hardness and yield stress should be combined with improved plasticity and high-temperature fracture strength for longer tool life. After annealing at 900°C the resultant AlTiN coating does not display an optimum balance of these properties. Annealing at 700°C gives the AlTiN coating optimum combination of properties for the interrupted cutting tests. It shows the best adaptive behavior with a combination of mechanical properties that provide a more stable environment for tribofilms to form and grow in interrupted cutting conditions than the coating after annealing at higher temperatures that display lower plasticity.

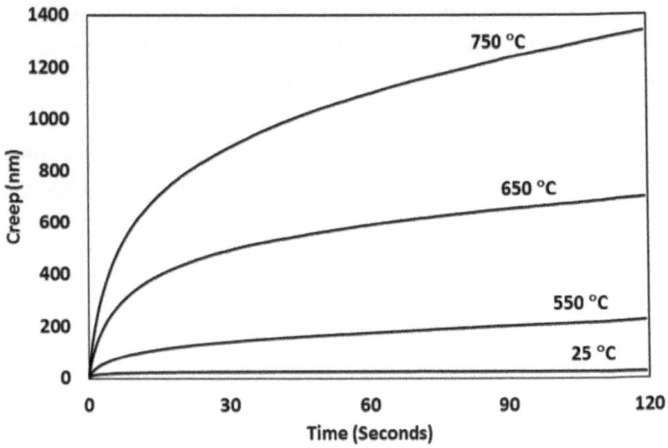

Figure 1.40 Influence of temperature on indentation creep of a glass–ceramic fuel cell seal material.

1.6.3.3 Fuel cell materials

Solid oxide fuel cells (SOFCs) are an increasingly popular high-efficiency form of energy conservation. Key to their performance and reliability is a hermetic seal that separates the fuel and air sides of the cell. SOFCs operate at around 800°C and at this temperature creep of the hermetic seal material can lead to early failure. The following example shows how the rate of deformation under a fixed load (creep) of a glass–ceramic fuel cell seal material increases with increasing temperature [130] (Fig. 1.40). High instrumental stability is required to undertake such measurements reliably.

In the same study, the glass–ceramic seal material was subjected to a thermal "preaging" process with the objective being to improve stability of high-temperature mechanical properties. Hardness and elastic modulus measurements were undertaken at 750°C. The measurements were carried out in an air environment and a cBN indenter was used as it would not oxidize at the testing temperature. One hundred hours of thermal aging result in the glass–ceramic exhibiting a smaller drop in elastic modulus with increasing temperature (Fig. 1.41). As such the aging process enhances the material's suitability for the SOFC seal application.

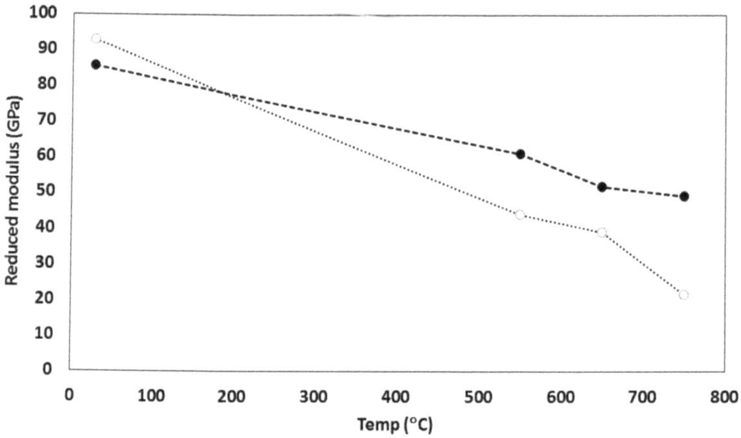

Figure 1.41 Influence of temperature on the elastic modulus of a glass–ceramic fuel cell seal material previously thermally aged at 800°C for 4 h (open circles) or 100 h (closed circles).

1.6.4 *Nonambient-Temperature Nanoscratch and Nanoimpact Testing*

1.6.4.1 High-temperature nanoscratch and friction testing

In many engineering applications frictional heating results in a local increase in the temperature between contacting surfaces and in high-speed dry machining temperatures may be 1000°C or more. For convenience macroscale scratch and wear testing of coatings is almost always performed at room temperature with unknown heating due to friction. A promising alternative approach is to perform mechanical and tribological tests with temperature control using heated stages and more moderate loads. Microscratch and microwear tests on wear-resistant coatings have been performed to 600°C using $R = 25$ μm diamond probes to apply loads of 0.5–20 N. He et al. investigated the low load sliding of silver–bismuth alloys from room temperature to 280°C [131]. Depending on the alloy composition they observed either (i) a sharp decrease in friction on melting or (ii) a general increase in friction with increasing temperature.

Diamond oxidizes in air above ~500°C, so the use of alternate indenter materials is required to extend the temperature range

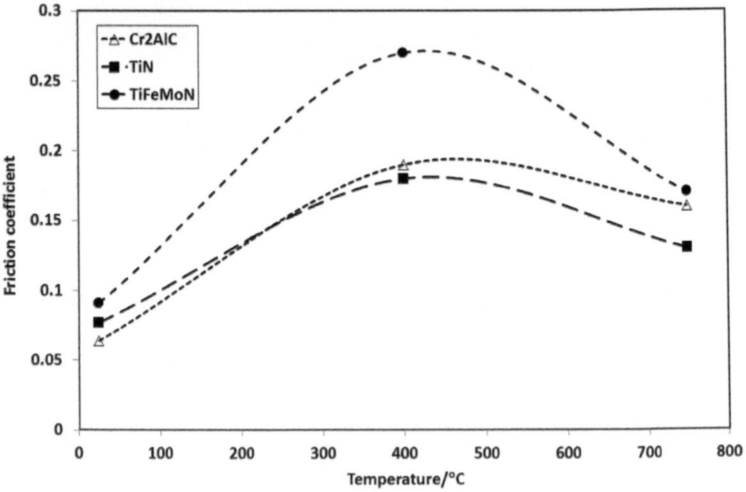

Figure 1.42 Variation in friction with temperature.

for applications such as high-temperature solid lubricants, TBCs, and cutting tool coatings that require higher temperatures. In a more recent study Smith et al. used a WC-Co probe to study nanoscale friction on a range of TiN-based coatings at 25°C, 400°C, and 750°C [132]. The tests were performed at very low contact pressure so that the ploughing contribution to the total measured friction, usually the dominant term at higher contact pressure, was minimized and the interfacial friction could be measured directly. A maximum in friction coefficient was observed at 400°C (Fig. 1.42). The decrease in friction at 750°C was associated with formation of lubricating surface oxides and oxidation-associated surface roughening.

In addition to the elevated-temperature nanoindentation, scratch, and friction testing, several other tests can be performed at temperatures of 500°C or more using commercial instrumentation (NanoTest system), as illustrated by Table 1.5 below.

Micropillar compression at elevated temperature is gaining popularity and is a relatively simple way of understanding how deformation mechanisms are affected by temperature [43, 133, 134]. Korte and Clegg [43] have reported micropillar compression

Table 1.5 Nanomechanical test techniques at elevated temperature

Technique	Applications	Maximum temperature (°C)
Nanoindentation	Fuel cells	750
Micropillar compression	Si, Ni-based superalloys	665
Microcantilever bending	W and Si beams	770
Nanofriction	TiN-based coatings	750
Microscratch and wear	Tool coatings	600
Nanoimpact	Tool coatings	500

Source: After Ref. [132].

of MgAl2O4 spinel at 25°C, 200°C, and 400°C using the NanoTest system with dual active heating of the diamond flat punch and the spinel sample. At 25°C only a small amount of plastic flow was observed before the pillars break. In contrast, pillars deformed entirely by plastic deformation at 200°C and 400°C. The measured yield stresses were consistent with those obtained previously by compressing larger pillars under high confining pressure, implying that the micropillar compression technique may be a useful way to assess plastic flow in brittle materials. Decreasing the sample size has reduced the temperature at which plasticity is normally observed by ~1500°C

As well as mechanical property determination, microcantilevers can be used for brittle to ductile transition studies. In this case microcantilevers are produced with a notch at their fixed end to act as a stress raiser for crack initiation. The cantilevers are then loaded at a known distance from their fixed end until a predetermined deformation is reached. Individual cantilevers are tested of the relevant temperature range until the transition from a ductile to brittle response is observed.

At 110°C the loading segment is seen to be essentially linear until brittle fracture occurs at about 0.5 μm displacement. At 700°C the loading is initially linear then the material starts to flow plastically, no fracture occurs. In both cases the experiments were configured to stop applying load once a displacement of 2 μm was reached (Fig. 1.43).

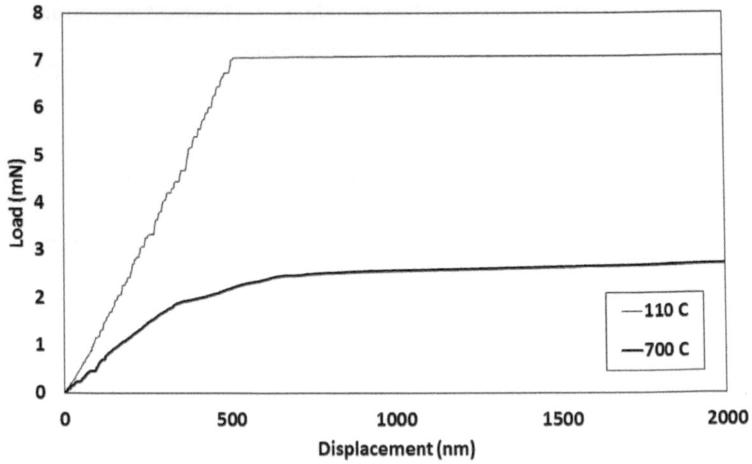

Figure 1.43 Load–displacement traces from bend tests on silicon cantilevers at 110°C and 700°C.

1.6.4.2 Subambient-temperature nanoindentation, nanoscratch, and nanowear testing

When testing mechanical properties below ambient temperature it is just as important that the sample and the indenter be at the same temperature as they are in elevated-temperature testing. Using dual-cooling control Chen et al. reported the development of a prototype cold-stage accessory for the NanoTest system, demonstrating its ability to investigate the local mechanical properties and creep behavior of atactic polypropylene down to −30°C [135]. The cooling system used incorporated a purging chamber to eliminate condensation during cooling and two Peltier coolers to achieve the isothermal contact. The subambient test capability has also been used to investigate the nanotribological and nanomechanical properties of a functionally graded a-C:H(Ti)/TiCN/TiN/Ti coating from 25°C to −30°C [136]. The coating failed by cracking and spallation in the room-temperature nanoscratch tests. Analysis of fracture sections showed these failures originated from, or close to, the interface between the top a-C:H(Ti) layer and the TiCN layer. Interestingly, decreasing the test temperature improved its properties. Testing at 0°C resulted in an increase in H, H/E_r, and H^3/E_r^2, consistent with improved crack resistance and wear resistance, with further modest improvements when the test temperature was reduced to –30°C.

1.6.4.3 High-temperature nanoimpact testing

Elevated-temperature nanoimpact tests have been performed for hard coatings and polymeric materials. Fracture probability data from nanoimpact tests at 500°C on TiAlN- and AlTiN-coated WC-Co inserts have been shown to correlate with the PI and cutting tool life data [137]. The nanoimpact test is sensitive to changes in the viscoelastic behavior of polymers as the test temperature is altered relative to the glass transition temperature. Constantinides et al. performed nanoimpact testing on polystyrene (PS) and polycarbonate (PC) from well below to well above their glass transition temperature ranges (testing from 20°C–180°C) [138]. They found that the coefficient of restitution e decreased only very slightly over the temperature range 0.2–1.0 T/T_g for both polymers. However, for temperatures exceeding the glass transition temperature (i.e., $T/T_g > 1$) the capacity of the materials to dissipate the energy of impact greatly (e decreases) increases.

1.7 Summary and Outlook

Technological developments over last few decades have introduced new and powerful techniques into the field of nanomechanical testing of materials, enabling the scale of experiments to be drastically decreased in terms of applied loads, relative displacements, and contact pressures. Improved resolution, calibration, and efficiency of existing techniques and development of new tools have allowed highly localized and accurate mechanical property information to be obtained.

Small-scale tribological experiments bring new opportunities to investigate the interaction between surfaces and help to build fundamental understanding of friction, lubrication, and wear of materials at the single asperity contact. This type of experiments requires specific, high-resolution equipment and methods with good stability and ultralow drift. The range of available nanomechanical testing techniques has drastically expanded beyond simple nanoindentation, including more complex mechanical testing scenarios, including nanoscratch and nanowear experiments. This has been driven by novel materials development and new application challenges but also by further expansion of commercial testing platforms capability.

As a result of expanded nanomechanical testing landscape, novel applications and methods have emerged. This includes improved tribological techniques as experiments at smaller length scales allow tribologists to build a fundamental understanding of frictional processes. Accelerated wear-testing and thin-film applications small-scale fretting experiments are some of these areas, as discussed earlier in this chapter.

Future challenges for nanomechanical testing require experiments in more severe test environments, including extreme low and high temperatures, various liquids, vacuum conditions, and the ability to handle increased energy densities. There will be also a constant pressure on equipment manufacturers to innovate with novel solutions allowing closer simulations of the real contact conditions in applications like MEMS and responding adequately to evolving miniaturization of mechanical systems and emerging areas such as bioengineering. Improved integration of measurement data obtained from mechanical testing seems to be increasingly a key part in providing reliable inputs for predictive surface damage models. Enhanced numerical models based on reliable mechanical data should not only allow better analysis of tribological experiments but also to forward-simulate such tests and provide hints for better component lifetime predictions.

References

1. ISO 14577: Metallic materials: instrumented indentation test for hardness and materials parameters, part 4: test method for metallic and non-metallic coatings.
2. Fischer-Cripps, A.J. (2011). *Nanoindentation*, 3rd Ed. (Springer, USA).
3. Fischer-Cripps, A.J. (2000). *Introduction to Contact Mechanics* (Springer, USA).

4. Beake, B.D., Bell, G.A., Goodes, S.R., Pickford N.J., and Smith, J.F. (2010). Improved nanomechanical test techniques for surface engineered materials, *Surf. Eng.*, **26**, 37.
5. Chudoba, T. (2006). Measurement of hardness and Young's modulus of coatings by nanoindentation, in *Nanostructured Coatings*, Cavaleiro, A., and De Hosson, J.Th.M., eds. (Springer, New York), 216–260.
6. Kermouche, G., Barthel, E., Vamdembroucq, D., and Dubujet, Ph. (2008). Mechanical modelling of indentation induced densification in amorphous silica, *Acta Mater.*, **56**, 3222–3228.
7. Beake, B.D., and Smith, J.F. (2002). High-temperature nanoindentation testing of fused silica and other materials, *Philos. Mag. A*, **82**, 2179–2186.
8. Chudoba, T., and Jennett, N.M. (2008). Higher accuracy analysis of instrumented indentation data obtained with pointed indenters, *J. Phys. D: Appl. Phys.*, **41**, 215407.
9. Morris, D.J., Myers, S.G., and Cook, R.F. (2004). Sharp probes of varying acuity: instrumented indentation and fracture behaviour, *J. Mater. Res.*, **19**, 165–175.
10. Gan, B., and Tin, S. (2010). Assessment of effectiveness of transition metal solutes in hardening of Ni solid solutions, *Mater. Sci. Eng. A*, **527**, 6809–6815.
11. Chudoba, T., and Richter, F. (2001). Investigation of creep behaviour under load during indentation experiments and its influence on hardness and modulus results, *Surf. Coat. Technol.*, **148**, 191–198.
12. Young, T.J., Monclus, M.A., Burnett, T.L., Broughton, W.R., Ogin, S.L., and Smith, P.A. (2011). The use of PeakForce^{TM} quantitative nanomechanical mapping AFM-based method for high-resolution Young's modulus measurement of polymers, *Meas. Sci. Technol.*, **22**, 125703.
13. Dub, S.N., Lim, Y.Y., and Chaudhri, M.M. (2010). Nanohardness of high purity Cu (111) single crystals: the effect of indenter load and prior plastic strain, *J. Appl. Phys.*, **107**, 043510.
14. Schwarzer, N., Duong, Q.-H., Bierwisch, N., Favaro, G., Fuchs, M., Kempe, P., Widrig, B., and Ramm, J. (2011). Optimisation of the scratch test for specific coating designs, *Surf. Coat. Technol.*, **206**, 1327–1335.
15. Van Vliet, K.J., Prchlik, L., and Smith, J.F. (2004). Direct measurement of indentation frame compliance, *J. Mater. Res.*, **19**, 325–331.
16. Moseson, A.J., Basu, S., and Barsoum, M.W. (2008). Determination of the effective zero point for spherical nanoindentation, *J. Mater. Res.*, **23**, 204–209.

17. Chiang, C.-J., Bull, S.J., Winscom, C., and Monkman, A. (2010). A nano-indentation study of the reduced elastic modulus of Alq3 and NPB thin-film used in OLED devices, *Org. Electron.*, **11**, 450–455.
18. Juliano, T., Gogotsi, Y., and Domnich, V. (2003). Effect of indentation unloading conditions on phase transformation induced events in silicon, *J. Mater. Res.*, **18**, 1192–1201.
19. Lu, C., Mai, Y.-W., Tam, P.L., and Shen, Y.G. (2007). Nanoindentation-induced elastic-plastic transition and size effect in α-Al2O3(0001), *Philos. Mag. Lett.*, **87**, 409–415.
20. Nix, W.D., and Gao, H. (1998). Indentation size effects in crystalline materials: a law for strain gradient plasticity, *J. Mech. Phys. Solids*, **46**, 411–425.
21. Qiao, X.G., Starink, M.J., and Gao, N. (2010). The influence of indenter tip rounding on the indentation size effect, *Acta Mater.*, **58**, 3690–3700.
22. Milman, Yu.V., Golubenko, A.A., and Dub, S.N. (2011). Indentation size effect in nanohardness, *Acta Mater.*, **59**, 7480–7487.
23. Cheng, Y.-T., and Cheng, C.-M. (1998). Relationships between hardness, elastic modulus, and the work of indentation, *Appl. Phys. Lett.*, **73**, 614.
24. Cheng, Y.-T., and Cheng, C.-M. (2004). Scaling, dimensional analysis, and indentation measurements, *Mater. Sci. Eng. R*, **44**, 91.
25. Greenwood, J., and Williamson, J. (1960). Contact of nominally flat surfaces, *Proc. R. Soc. A*, **295**, 300.
26. Beake, B.D., Fox-Rabinovich, G.S., Veldhuis S.C., and Goodes, S.R. (2009). Coating optimisation for high-speed machining with advanced nanomechanical test methods, *Surf. Coat. Technol.*, **203**, 1919–1925.
27. Beake, B.D, Goodes, S.R., Smith, J.F., Fox-Rabinovich, G.S., and Veldhuis, S.C. (2010). Using nanomechanics to optimise coatings for cutting tools, in *Handbook of Nanostructured Thin Films and Coatings, Mechanical Properties*, Zhang, S., ed. (CRC Press), 205–244.
28. Leyland, A., and Matthews, A. (2000). On the significance of the H/E ratio in wear control: a nanocomposite coating approach to optimised tribological behavior, *Wear*, **246**, 1–11.
29. Leyland, A., and Matthews, A. (2004). Design criteria for wear-resistant nanostructured and glassy-metal coatings, *Surf. Coat. Technol.*, **177–178**, 317.
30. Beake, B.D., Vishnyakov, V.M., Valizadeh, R., and Colligon, J.S. (2006). Influence of mechanical properties on the nanoscratch behaviour of hard nanocomposite TiN/Si3N4 coatings on Si, *J. Phys. D: Appl. Phys.*, **39**, 1392.

31. Bull, S.J. (2006). Using work of indentation to predict erosion behaviour in bulk materials and coatings, *J. Phys. D: Appl. Phys.*, **39**, 1624.

32. Beake, B.D., Vishnyakov, V.M., and Colligon, J.S. (2011). Nano-impact testing of TiFeN and TiFeMoN films for dynamic toughness evaluation, *J. Phys. D: Appl. Phys.*, **44**, 085301.

33. Korunsky, A.M., and Constantinescu, A. (2006). Work of indentation approach to the analysis of hardness and modulus of thin coatings, *Mater. Sci. Eng. A*, **423**, 28–35.

34. Veprek, S., Niederhofer, A., Moto, K., Bolom, T., Männling, H-D., Nesladek, P., Dollinger, G., and Bergmaier, A. (2000). Composition, nanostructure and origin of the ultrahardness in nc-TiN/a-Si_3N_4 and nc-$TiSi_2$ nanocomposites with H_v = 80 GPa to $\geq$ 105 GPa, *Surf. Coat. Technol.*, **133–134**, 152–159.

35. Fischer-Cripps, A.C., Bull, S.J., and Schwarzer, N. (2012). Critical review of claims for ultra-hardness in nanocomposite coatings, *Philos. Mag.*, **92**, 1601–1630.

36. Cordill M.J., Lund, M.S., Parker, J., Leighton, C., Nair, A.K., Farkas, D., Moody, N.R., and Gerberich, W.W. (2009). The nano-jackhammer effect in probing near-surface mechanical properties, *Int. J. Plasticity*, **25**, 2045–2058.

37. Cook, R.F., and Pharr, G.H. (1990). Direct observation and analysis of indentation cracking in glasses and ceramics, *J. Am. Ceram. Soc.*, **73**, 787.

38. Chen, J. (2012). Indentation-based methods to assess fracture toughness for thin coatings, *J. Phys. D: Appl. Phys.*, **45**, 203001.

39. Anstis, G.R., Chantikul, P., Lawn, B.R., and Marshall, D.B. (1981). A critical evaluation of indentation techniques for measuring fracture toughness: I. Direct crack measurements, *J. Am. Ceram. Soc.*, **64**, 533.

40. Chen, J., and Bull, S.J. (2007). Indentation fracture and toughness assessment for thin optical coatings on glass, *J. Phys. D: Appl. Phys.*, **40**, 5401.

41. Uchic, M.D., Dimiduk, D.M., Florando, J.N., and Nix, W.D. (2004). Sample dimensions influence strength and crystal plasticity, *Science*, **305**, 986.

42. Greer, J.R., Oliver, W.C., and Nix, W.D. (2005). Size dependence of mechanical properties of gold at the micron scale in the absence of strain gradients, *Acta Mater.*, **53**, 1821.

43. Korte, S.K., and Clegg, W.J. (2009). Micropillar compression of ceramics at elevated temperature, *Scr. Mater.*, **60**, 807.

44. Jennett, N.M., Ghisleni, R., and Michler, J. (2009). Enhanced yield strength of materials: the thinness effect, *Appl. Phys. Lett.*, **95**, 123102.
45. Armstrong, D.E.J., Wilkinson, A.J., and Roberts, S.G. (2009). Measuring anisotropy in Young's modulus of copper using microcantilever testing, *J. Mater. Res.*, **24**, 3268.
46. Rollin, M., Jouannigot, S., Lamon, J., and Pailler, R. (2009). Characterization of fibre/matrix interfaces in carbon/carbon composites, *Compos. Sci. Technol.*, **69**, 1442–1446.
47. Steinmann, P.A., Tardy, Y., and Hintermann, H.E. (1987). Adhesion testing by the scratch test method: the influence of intrinsic and extrinsic parameters on the critical load, *Thin Solid Films*, **154**, 333.
48. Beake, B.D., Ogwu, A.A., and Wagner, T. (2006). Influence of experimental factors and film thickness on the measured critical load in the nanoscratch test, *Mater. Sci. Eng. A*, **423**, 70–73.
49. Beake, B.D., Davies, M.I., Liskiewicz, T.W., Vishnyakov, V.M., and Goodes S.R. (2013). Nano-scratch, nanoindentation and fretting tests of 5–80 nm ta-C films on Si(100), *Wear*, **301**, 575–582.
50. Wu, T.W. (1991). Microscratch and load relaxation tests for ultra-thin films, *J. Mater. Res.*, **6**, 407.
51. Beake, B.D., Vishnyakov, V.M., and Harris, A.J. (2011). Relationship between mechanical properties of thin nitride-based films and their behaviour in nano-scratch tests, *Tribol. Int.*, **44**, 468–475.
52. Klocke, F., Bouzakis, K.-D., Georgiadis, K., Geradis. S., Skordaris, G., and Pappa, M. (2011). Adhesive interlayers' effect on the entire structure strength of glass modling tools' Pt-Ir coatings by nano-tests determined, *Surf. Coat. Technol.*, **206**, 1867–1872.
53. Beake, B.D., Shi, B., and Sullivan, J.L. (2011). Nanoscratch and nanowear testing of TiN coatings on M42 steel, *Tribology*, **5**, 141–147.
54. Beake, B.D., Goodes, S.R., and Shi, B. (2009). Nanomechanical and nanotribological testing of ultra-thin carbon-based and MoST films for increased MEMS durability, *J. Phys. D: Appl. Phys.*, **42**, 065301.
55. Beake, B.D., Liskiewicz, T.W., and Smith, J.F. (2011). Deformation of Si(100) in spherical contacts: comparison of nano-fretting and nano-scratch tests with nano-indentation, *Surf. Coat. Technol.*, **206**, 1921–1926.
56. Beake, B.D., and Lau, S.P. (2005). Nanotribological and nanomechanical properties of 5-80 nm tetrahedral amorphous carbon films on silicon, *Diam. Relat. Mater.*, **14**, 1535–1542.

57. Shi, B., Sullivan, J.L., and Beake, B.D. (2008). An investigation into which factors control the nanotribological behaviour of thin sputtered carbon films, *J. Phys. D: Appl. Phys.*, **41**, 045303.
58. Bull, S.J., and Rickerby, D.S. (1989). Multi-pass scratch testing as a model for abrasive wear, *Thin Solid Films*, **181**, 545.
59. Beake, B.D., and Ranganathan, N. (2006). An investigation of the nanoindentation and nano/micro-tribological behaviour of monolayer, bilayer and trilayer coatings on cemented carbide, *Mater. Sci. Eng. A*, **423**, 46.
60. Collins, J.A. (1981). *Failure of Materials in Mechanical Design* (John Wiley, New York).
61. Sato K., Stolarski, T.A., and Iida, Y. (2000). The effect of magnetic field on fretting wear, *Wear*, **241**, 99–108.
62. Mohrbacher, H., Celis, J.-P., and Roos, J.R. (1995). Laboratory testing of displacement and load induced fretting, *Tribol. Int.*, **28**, 269–278.
63. Halliday, J.S., and Hirst, W. (1956). The fretting corrosion of mild steel, *Proc. R. Soc. A*, **236**, 411–425.
64. Vaessen, G.H.G., Commissaris, C.P.L., and De Gee, A.W.J. (1968–1969). Fretting corrosion of Cu-Ni-Al against plain carbon steel, *Proc. Inst. Mech. Eng.*, **183**, 125.
65. Ohmae, N., and Tzukizoe T. (1974). The effect of slip amplitude on fretting, *Wear*, **27**, 281–293.
66. Gordelier, S.C., and Chivers, T.C. (1979). A literature review of palliatives for fretting fatigue, *Wear*, **56**, 177–190.
67. Gordelieen, S.C., and Stallings, L. (1982). An evaluation of fretting at small slip amplitudes, *ASTM Spec. Tech. Publ.*, **780**, 30–48.
68. Kennedy, P.J., Stallings, L., and Petersen, M.B. (1983). A study of surface damage at low amplitude slip, *Proc. Conf. "ASLE-ASME Lubrication Conf."*, Hartford.
69. Fouvry, S., Kapsa, Ph., and Vincent, L. (1996). Quantification of fretting damage, *Wear*, **200**, 186–205.
70. Williams, J., and, Le, H. (2006). Tribology and MEMS, *J. Phys. D*, **39D**, R201.
71. Drees, D., Celis, J.-P., and Achanta, S. (2004). Friction of thin coatings on three length scales under reciprocating sliding, *Surf. Coat. Technol.*, **188–189**, 511–518.
72. Varenberg, M., Etsion, I., and Halperin, G. (2005). Nanoscale fretting wear study by scanning probe microscopy, *Tribol. Lett.*, **18**(4), 493–498.

73. Yoon, Y., Etsion, I., and Talke, F.E. (2011). The evolution of fretting wear in a micro-spherical contact, *Wear*, **270**, 567–575.
74. Yu, J.X., Qian, L.M., Yu, B.J., and Zhou, Z.R. (2009). Nanofretting behaviours of monocrystalline silicon (100) against diamond tips in atmosphere and vacuum, *Wear*, **267**, 322–329.
75. Yu, J., Qian, L., Yu, B., and Zhou, Z. (2009). Nanofretting behaviour of monocrystalline silicon (100) against SiO2 microsphere in vacuum, *Tribol. Lett.*, **34**, 31–40.
76. Chen, L., Yang, M., Yu, J., Qian, L., and Zhou, Z. (2011). Nanofretting behaviours of ultrathin DLC coating on Si(100) substrate, *Wear*, **271**, 1980–1986.
77. Wilson, G.M., Smith, J.F., and Sullivan, J.L. (2008). A nanotribological study of thin amorphous C and Cr doped amorphous C coatings, *Wear*, **265**, 1633–1641.
78. Wilson, G.M., Smith, J.F., and Sullivan, J.L. (2009). A DOE nano-tribological study of thin amorphous carbon-based films, *Tribol. Int.*, **42**, 220–228.
79. Wilson, G.M., and Sullivan, J.L. (2009). An investigation into the effect of film thickness on nanowear with amorphous carbon-based coatings, *Wear*, **266**, 1039–1043.
80. Achanta, S., Drees, D., and Celis, J.-P. (2005). Friction and nanowear of hard coatings in reciprocating sliding at milli-Newton loads, *Wear*, **259**, 719–729.
81. Achanta, S., Drees, D., and Celis J.-P. (2007). Investigation of friction on hard homogeneous coatings during reciprocatings tests at micro-Newton normal forces, *Wear*, **263**, 1390–1396.
82. Weppelmann, E.R., Field J.S., and Swain, M.V. (1995). Influence of spherical indentor radius on the indentation-induced transformation behavior of silicon, *J. Mater. Sci.*, **30**, 2455.
83. Bradby, J.E., Williams, J.S., Wong-Leung, J., Swain, M.V., and Munroe, P. (2001). Mechanical deformation in silicon by micro-indentation, *J. Mater. Res.*, **16**, 1500–1507.
84. Juliano, T., Domnich, V., and Gogotsi, Y. (2004). Examining pressure-induced phase transformations in silicon by spherical indentation and Raman spectroscopy: a statistical study, *J. Mater. Res.*, **19**, 3099–3108.
85. Chang, L., and Zhang, L. (2009). Deformation mechanisms at pop-out in monocrystalline silicon under nanoindentation, *Acta Mater.*, **57**, 2148–2153.

86. Mitchell, A., and Shrotriya, P. (2007). Onset of nanoscale wear of metallic implant materials: influence of surface residual stresses and contact loads, *Wear*, **263**, 1117–1123.
87. Corbella, C., Rubio-Roy, M., Bertran, E., Polo, M.C., Pascual, E., and Andújar, J.L. (2009). Low friction and protective diamond-like carbon coatings deposited by asymmetric bipolar pulsed plasma, *Diam. Relat. Mater.*, **18**, 1035–1038.
88. Liskiewicz, T.W., Beake, B.D., and Smith, J.F. (2010). In situ accelerated micro-wear: a new technique to fill the measurement gap, *Surf. Coat. Technol.*, **205**, 1455–1459.
89. Ku, I.S.Y., Reddyhoff, T., Wayte, R., Choo, J.H., Holmes, A.S., and Spikes, H.A. (2012). Lubrication of microelectromechanical devices using liquids of different viscosities, *J. Tribol.*, **134**, 012002-1.
90. Maboudian, R., Ashhurst, W.R., and Carraro, C. (2000). Self-assembled monolayers as anti-stiction coatings for MEMS: characterisation and recent developments, *Sens. Actuators*, **82**, 219–223.
91. Smallwood, S.A., Eapen, K.C., Patton, S.T., and Zabinski, J.S. (2006). Performance results of MEMS coated with a conformal DLC, *Wear*, **260**, 1179–1189.
92. Akita, N., Konishi, Y., Ogura, S., Imamura, M., Hu, Y.H., and Shi, X. (2001). Comparison of deposition methods for ultra thin DLC overcoat film for MR head, *Diam. Relat. Mater.*, **10**, 1017–1023.
93. Chan, C.Y., Lai, K.H., Fung, M.K., Wong, W.K., Bello, I., Huang, R.F., Lee, C.S., Lee, S.T., and Wong, S.P. (2001). Deposition and properties of tetrahedral amorphous carbon films prepared on magnetic disks, *J. Vac. Sci. Technol. A*, **19**, 1606.
94. Wang, F.C., Jin, Z.M., McEwen, H.M., and Fisher, J. (2003). Microscopic asperity contact and deformation of ultrahigh molecular weight polyethylene bearing surfaces, *Proc. Inst. Mech. Eng. H*, **217**, 477–490.
95. Raimondi, M.T., Vena, P., and Pietrabissa, R. (2001). Quantitative evaluation of the prosthetic head damage induced by microscopic third-body particles in total hip replacement, *J. Biomed. Mater. Res. B*, **58**, 436–448.
96. Beake, B.D., and Liskiewicz, T.W. (2013). Comparison of nano-fretting and nano-scratch tests on biomedical materials, *Tribol. Int.*, **63**, 123–131.
97. Fox-Rabinovich, G.S., Yamamoto, K., Beake, B.D., Kovalev, A.I., Aguirre, M.H., Veldhuis, S.C., Dosbaeva, G., Wainstein, D.L., Biksa, A., and Rashkovskiy, A.Y. (2010). Emergent behavior of nano-multilayered

coatings during dry high speed machining of hardened tool steels, *Surf. Coat. Technol.*, **204**, 3425.

98. Fox-Rabinovich, G.S., Beake, B.D., Veldhuis, S.C., Endrino, J.L., Parkinson, R., Shuster, L.S., and Migranov, M.S. (2006). Impact of mechanical properties measured at room and elevated temperatures on wear resistance of cutting tools with TiAlN and AlCrN coatings, *Surf. Coat. Technol.*, **200**, 5738.
99. Chen, J., Beake, B.D., Wellman, R., Nicholls, J.R., and Dong, H. (2012). An investigation into the correlation between nano-impact resistance and erosion performance of EB-PVD thermal barrier coatings on thermal ageing, *Surf. Coat. Technol.*, **206**, 4992–4998.
100. Trelewicz, J.R., and Schuh, C.A. (2008). The Hall–Petch breakdown at high strain rates: optimizing nanocrystalline grain size for impact applications, *Appl. Phys. Lett.*, **93**, 171916.
101. Constantinides, G., Tweedie, C.A., Savva, N., Smith, J.F., and Van Vliet, K.J. (2009). Quantitative impact testing of energy dissipation at surfaces, *Exp. Mech.*, **49**, 511–522.
102. Beake, B.D., and Smith, J.F. (2004). Nano-impact testing: an effective tool for assessing the resistance of advanced wear-resistant coatings to fatigue failure and delamination, *Surf. Coat. Technol.*, **188–189C**, 594.
103. Jennett, N.M., and Nunn, J. (2011). High resolution measurement of dynamic (nano) indentation impact energy: a step towards the determination of indentation fracture resistance, *Philos. Mag.*, **91**, 1200–1220.
104. Bouzakis, K.-D., Gerardis, S., Skordaris, G., and Bouzakis, E. (2011). Nano-impact tests on a TiAlN PVD coating and correlation between experimental and FEM results, *Surf. Coat. Technol.*, **206**, 1936–1940.
105. Faisal, N.H., Ahmed, R., and Fu, Y. (2011). Nano-impact testing and failure mechanism of solar panel DLC film, *Proc. ATMC 2011, KACST,* Saudi Arabia, 36–44.
106. Faisal, N.H., Ahmed, R., and Fu, Y. (2012). Nano-impact (fatigue) characterisation of as-deposited amorphous nitinol thin film, *Coatings*, **2**, 195–209.
107. Bouzakis, K.-D., Skordaris, G., Bouzakis, E., Tsouknidas, A., Makrimallakis, S., Gerardis, S., and Katirtzoglou, G. (2011). Optimisation of wet micro-blasting on PVD films with various grain materials for improving the coated tools' cutting performance, *CIRP Ann.: Manuf. Technol.*, **60**, 587–590.

108. Bouzakis, K.-D., Klocke, F., Skordaris, G., et al. (2011). Influence of dry micro-blasting grain quality on wear behaviour of TiAlN coated tools, *Wear*, **271**, 783–791.

109. Bouzakis, K.-D., Bouzakis, E., Skordaris, G., et al. (2011). Effect of PVD films wet micro-blasting by various Al2O3 grain sizes on the wear behaviour of coated tools, *Surf. Coat. Technol.*, **205**, S128–S132.

110. Chen, J., Ji, R., Khan, R.H.U., Li, X., Beake B.D., and Dong, H. (2011). Effects of mechanical properties and layer structure on the cyclic loading of TiN-based coatings, *Surf. Coat. Technol.*, **206**, 522.

111. Altaf, K., Ashcroft, I.A., and Hague, R. (2012). Modelling the effect of moisture on the depth sensing indentation response of a stereolithography polymer, *Comput. Mater. Sci.*, **52**, 112–117.

112. Bushby, A.J., Ferguson, V.L., and Boyde, A. (2004). Nanoindentation of bone: comparison of specimens tested in liquid and embedded in polymethylmethacrylate, *J. Mater. Res.*, **19**, 249–259.

113. Bell, G.A., Bieliński D.M., and Beake, B.D. (2008). Influence of water on the nanoindentation creep response of nylon 6, *J. Appl. Polym. Sci.*, **107**, 577–582.

114. Constantinides, G., Kalcioglu, Z.I., McFarland, M., et al. (2008). Probing mechanical properties of fully hydrated gels and biological tissues, *J. Biomech.*, **41**, 3285–3289.

115. Beake, B.D., Chen, J., Gao, F., Harris, A.J., and Bell, G.A. (2015). Influence of humidity and temperature on the nanomechanical behaviour of polymeric materials, in preparation.

116. Schmidt, D.J., Cebeci, F.C., Kalcioglu, Z.I., et al. (2009). Electrochemically controlled swelling and mechanical properties of a polymer nanocomposite, *ACS Nano*, **3**, 2207–2216.

117. Sun, D., Wharton, J.A., and Wood, R.J.K. (2009). Micro-abrasion mechanisms of cast CoCrMo in simulated body fluids, *Wear*, **267**, 1845–1855.

118. Kalcioglu, Z.I., Qu, M., Strawhecker, K.E., Shazly, T., Edelman, E., VanLandingham, M.R., Smith, J.F., and Van Vliet, K.J. (2010). Dynamic impact indentation of hydrated biological tissues and tissue surrogate gels, *Philos. Mag.*, **91**, 1339–1355.

119. Everitt, N.M., Davies, M.J., and Smith, J.F. (2011). High temperature nanoindentation: the importance of isothermal contact, *Philos. Mag.*, **91**, 1221–1244.

120. Tang, F., and Zhang, L.C. (2011). On the reliability of nanoindentation on Si wafer at elevated temperatures, *Adv. Mater. Res.*, **325**, 684.

121. Zarnetta, R., Kneip, S., Somsen, Ch., and Ludwig, A. (2011). High-throughput characterization of mechanical properties of Ti-Ni-Cu shape memory thin films at elevated temperature, *Mater. Sci. Eng. A*, **528**, 6552–6557.

122. Schuh, C.A., Packard, C.E., and Lund, A.C. (2006). Nanoindentation and contact-mode imaging at high temperatures, *J. Mater. Res.*, **21**, 725–736.

123. Trenkle, J.C., Packard, C.E., and Schuh, C.A. (2010). Hot nanoindentation in inert environments, *Rev. Sci. Instrum.*, **81**, 073901.

124. Franke, O., Trenkle, J.C., and Schuh, C.A. (2010). Temperature dependence of the indentation size effect, *J. Mater. Res.*, **25**, 1225–1229.

125. Smith, J.F., and Zheng, S. (2000). High temperature nanoscale mechanical property measurements, *Surf. Eng.*, **6**, 143.

126. UK Patent No. GB2478134B and Int. Pat. Appl. WO2011/104529.

127. Fox-Rabinovich, G.S., Endrino, J.L., Beake, B.D., Kovalev, A.I., Veldhuis, S.C., Ning, L., Fotaine, F., and Gray, A. (2006). Impact of annealing on the microstructure, properties, and cutting performance of AlTiN coating, *Surf. Coat. Technol.*, **201**, 3524.

128. Fox-Rabinovich, G.S. Endrino, J.L. Beake, B.D. Aguirre, M.H. Veldhuis, S.C Quinto, D.H., Bauer, C.E., Kovalev, A.I., and Gray, A. (2008). Effect of annealing below 900°C on structure, properties and tool life of an AlTiN coating under various cutting conditions, *Surf. Coat. Technol.*, **202**, 2985.

129. Bouzakis, K.-D., Pappa, M., Geradis. S., Skordaris, G., and Bouzakis, E. (2012). PVD coatings' strength at various temperatures by nanoindentations and FEM calculations determined, *Tribol. Ind.*, **34**, 29–35.

130. Milhans, J., Li, D.S., Khaleel, M., Sun, X. Al-Haik, M.S., Harris, A., and Garmenstani, H. (2011). Mechanical properties of solid oxide fuel cell glass-ceramic seal at high temperatures, *J. Power Sources*, **196**, 5599.

131. He, B., Ghosh, G., Chung, Y.-W., and Wang, Q. (2010). Effect of melting and microstructure on the microscale friction of silver–bismuth alloys, *Tribol. Lett.*, **38**, 275–282.

132. Smith, J.F., Vishnyakov, V.M., Davies, M.I., and Beake, B.D. (2013). Nanoscale friction measurements up to 750°C, *Tribol. Lett.*, **49**, 455–463.

133. Korte, S.K., Banard, J.S., Stearn, R.J., and Clegg, W.J. (2011). Deformation of silicon, insights from microcompression testing at 25–500°C, *Int. J. Plasticity*, **27**, 1853.

134. Korte, S.K., Stearn, R.J., Wheeler, J.M., and Clegg, W.J. (2012). High temperature microcompression and nanoindentation in vacuum, *J. Mater. Res.*, **27**, 167–176.

135. Chen, J., Bell, G.A., Dong, H., Smith, J.F., and Beake, B.D. (2010). A study of low temperature mechanical properties and creep behaviour of polypropylene using a new sub-ambient temperature nanoindentation test platform, *J. Phys. D: Appl. Phys.*, **43**, 425404.

136. Chen, J. Bell, G.A., Beake B.D., and Dong, H. (2011). Low temperature nano-tribological study on a functionally graded tribological coating using nanoscratch tests, *Tribol. Lett.*, **43**, 351.

137. Beake, B.D., Smith, J.F., Gray, A, Fox-Rabinovich, G.S., Veldhuis, S.C., and Endrino, J.L. (2007). Investigating the correlation between nano-impact fracture resistance and hardness/modulus ratio from nanoindentation at 25–500°C and the fracture resistance and lifetime of cutting tools with Ti1-xAlxN (x = 0.5 and 0.67) PVD coatings in milling operations, *Surf. Coat. Technol.*, **201**, 4585.

138. Constantinides, G., Tweedie, C.A., Holbrook, D.M., Barragan, P., Smith, J.F., and Van Vliet, K.J. (2008). Quantifying deformation and energy dissipation of polymeric surfaces under localized impact, *Mater. Sci. Eng. A*, **489**, 403–412.

Chapter 2

Surface Properties of Polymers and Rubber Measured by Nanoindentation

Dariusz M. Bieliński
Institute of Polymer and Dye Technology, Faculty of Chemistry, Łódź University of Technology, Stefanowskiego 12/16, 90-924 Łódź, Poland
Division of Elastomers and Rubber Technology, Institute for Engineering of Polymer Materials and Dyes, Harcerska 30, 05-820 Piastów, Poland
dbielin@p.lodz.pl

2.1 Introduction

Energetic, thermal, and chemical characteristics of the surface differ significantly from the bulk of polymer materials [1]. It is the result of their different chemical composition and physical structure, constituting already at the stage of material processing [2]. Tailoring properties of polymer and rubber goods is very often realized by chemical or physical treatment of their surface [3, 4]. Modification of the surface layer of polymer materials takes place also during their exploitation and accompanies aging processes [5]. Due to its limited depth, the role of the surface layer in polymer technology and applications used to be very often underestimated or even neglected. It is reflected, for example, by testing methods, dominated by determination of volume properties of materials and

Materials Characterization: Modern Methods and Applications
Edited by Narayanaswami Ranganathan

ISBN 978-981-4613-06-4 (Hardcover), 978-981-4613-07-1 (eBook)
www.panstanford.com

technical products, whereas many phenomena and processes such as abrasion, swelling, aging, or burning, deciding the most important properties of polymer materials and rubber, originate and continue in their surface layer.

Recent achievements in surface analysis have made it possible to detect and even to monitor changes taking place in the top layer of polymer materials in different stages of their life cycle. One of the achievements is undoubtedly the technique called nanoindentation, enabling the determination of mechanical properties: modulus and hardness of the surface layer, thin films, or polymer coatings [6, 7]. The advantage of nanoindentation over conventional methods of hardness testing lies in the fact of the measurement of real tensions, containing also the resilient component of material deformation, which makes it possible to study even very thin layers of materials. These layers are not accessible for conventional techniques, requiring plastic deformation of materials.

This chapter presents some examples of applications of the nanoindentation technique for studying:

- surface gradient of vulcanizate crosslinking;
- surface migration of low-molecular-weight components of rubber;
- surface segregation in polymer blends;
- chemical or physical surface treatment in order to obtain the expected degree of polymer modification;
- conditions of polyester resin photopolymerization from the point of view of exploitation properties of the polymer coatings; and
- monitoring of polymer aging or estimation of the resistance of polymer materials to swelling in fuels.

Existence of the surface gradient of vulcanizate crosslinking is demonstrated, explaining its origin and consequences. The methodology of low molecular weight bloom thickness determination by application of nanoindentation is presented. It has been proved that apart from migration of low-molecular-weight additions, it is also possible for surface segregation of the low-molecular-weight part of polymer components to take place in polymer mixes.

Also the influence of swelling in fuels and thermal or ozone aging on hardness, stress relaxation, and creep phenomena in the surface layer of elastomers and rubber is presented in the chapter. In the examples of polyester resin photopolymerization, chlorination, low-energy ion beam bombardment, and ultraviolet (UV) radiation treatment of rubber surface, one demonstrated how the progress of resin photocuring or physical and chemical modification of materials adequately could be monitored by determination of the mechanical properties of their surface layer.

2.2 Experiments

2.2.1 *Instrumentation*

Nanoindentation studies of materials were carried out with a NanoTest 600 instrument (Micro Materials Ltd., UK), equipped with a diamond indenter of Berkovitch geometry. The scheme, explaining the concept of its operation, is presented in Fig. 2.1 [6].

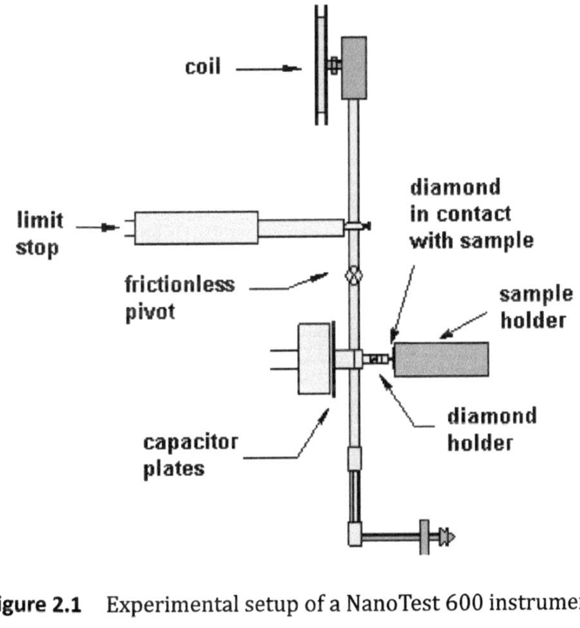

Figure 2.1 Experimental setup of a NanoTest 600 instrument.

2.2.2 *Methodology*

2.2.2.1 Nanoindentation

Figure 2.2 presents a theoretical curve representing the loading–unloading indentation cycle (limit stop at constant force), which enables calculation of hardness H and the so-called reduced Young modulus E_r (related to the real mechanical modulus) of the polymer tested at the depth h from the surface of the material.

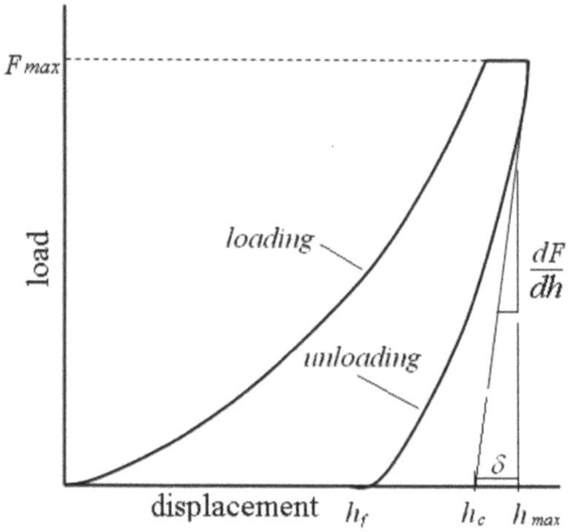

Figure 2.2 Schedule of the nanoindentation experiment.

The maximum depth of penetration h_{max}, adequate to the maximum loading force F_{max}, contains both the plastic component h_f (adequate to the unloading force equal to zero) as well as the resilient one δ.

The so-called contact depth h_c, for which the area of indenter–material contact, A, is determined, is used for the calculation of material hardness, according to Eq. 2.1:

$$H = F_{max}/A \qquad (2.1)$$

According to Oliver and Pharr [8], the contact depth h_c can be calculated according to Eq. 2.2:

$$h_c = h_{max} - 0.75\ F_{max}/S \qquad (2.2)$$

where S is a contact stiffness, represented by the angle of incidence of the slope to the unloading curve at its beginning period (Eq. 2.3):

$$S = \mathrm{d}F/\mathrm{d}h \tag{2.3}$$

The reduced Young modulus of the studied pair steel–rubber E_r can be calculated using Eq. 2.4:

$$E_\mathrm{r} = 0.5(\pi S/A)0.5 \tag{2.4}$$

Knowing the value of the elasticity modulus and the Poisson ratio for the material of indenter and assuming the Poisson ratio of rubber equal to $\nu = 0.5$, it is possible to calculate the elasticity modulus of the rubber studied at the defined depth, applying the well-known Hertz equation.

2.2.2.2 Creep

Monitoring the time penetration of the indenter loaded with a constant force into the material studied, one can examine kinetics of the creep of polymer materials and on this basis estimate the degree of modification of their surface layer. The best fit of experimental results to a theoretical model was obtained for the logarithmic dependence (Eq. 2.5), originally developed for soft metals [9–11]:

$$\Delta h = A\ln(Bt + 1) \tag{2.5}$$

where Δh is the depth of indenter penetration from the surface of the material studied, A and B are creep parameters, and t is time. Especially important is the parameter A, which, after normalization to the initial depth of penetration A/h_0, has been proposed as a dimensionless creep parameter [12], given by Eq. 2.6:

$$h/h_0 = (A/h_0)\ln(Bt + 1) \tag{2.6}$$

where h_0 is the material deformation at the beginning of the creep experiment under the maximum load, whereas h/h_0 is a relative increase of the deformation afterward.

2.2.2.3 Bloom thickness

Nanoindentation can be used for estimation of the thickness of low-molecular-weight components, segregating onto the surface of polymer mixes, blends, or composites. This time most adequate is a load–partial unload technique, as illustrated by the example of the bloom of stearic acid present on the surface of the styrene–butadiene rubber (SBR) vulcanizate (Fig. 2.3) [13].

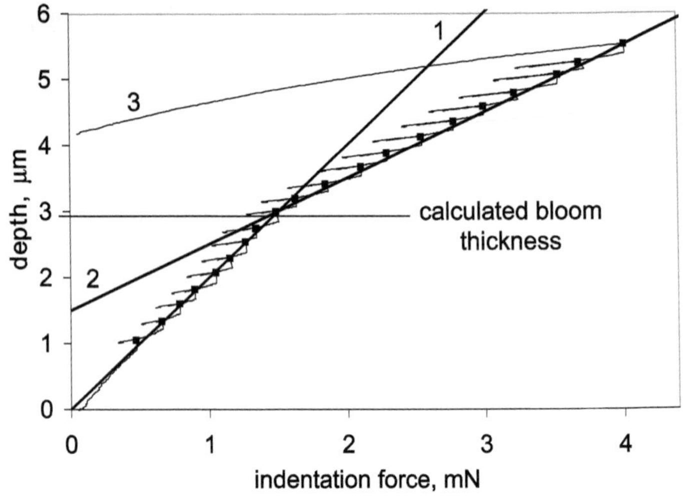

Figure 2.3 Methodology for the determination of low-molecular-weight substance bloom thickness, present on the surface of an SBR vulcanizate. (1, 2) Tangent to the load curve for the bloom region and for the polymer base adequately and (3) unloading curve.

The depth at which the slope of the experimental loading curve changes (adequately to the change of material stiffness) can be assigned to the indenter passage from the soft bloom to the polymer base of higher stiffness.

2.3 Results and Discussion

2.3.1 *Surface Gradient of Crosslinks in Rubber Vulcanizates*

Due to the limited thermal conductivity of polymers, during steel-mold vulcanization of big rubber parts the temperature in their surface layer is higher than inside, which influences both the crosslink density and the crosslink structure created in rubber vulcanizates. This is, first of all, reflected by their surface gradient of hardness, which influences the exploitation properties of rubber goods. It has been illustrated in the example of isoprene rubber (IR) crosslinked with dicumyl peroxide (DCP) (Fig. 2.4) [14].

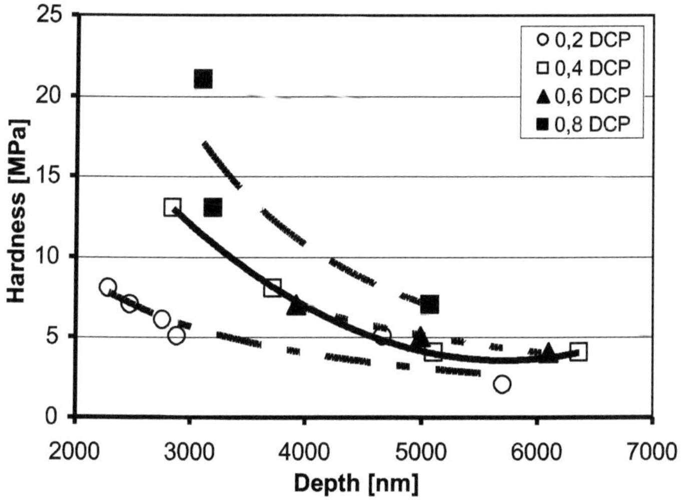

Figure 2.4 Hardness profile of isoprene rubber crosslinked with dicumyl peroxide (DCP). Numbers indicate the amount of DCP used for crosslinking (phr).

From Fig. 2.5 one can easily come to the conclusion that the gradient of mold temperature manifests itself by the gradient of crosslink density [15], which is reflected by the gradient of mechanical properties, which in turn has been confirmed by the results of studying rubber vulcanizates by nanoindentation.

In the case of sulfur vulcanization, the surface gradient of hardness is additionally influenced by the phenomenon of network maturing, which can be described as the destruction of polysulfide crosslinks, accompanied simultaneously by the creation of a higher amount of short mono- and disulfide ones instead [16]. The effect is pronounced to the highest extent for rubber crosslinked with "sulfur" systems—containing sulfur and accelerators (S + MBT or even better S + DPG, producing a significant amount of polysulfide

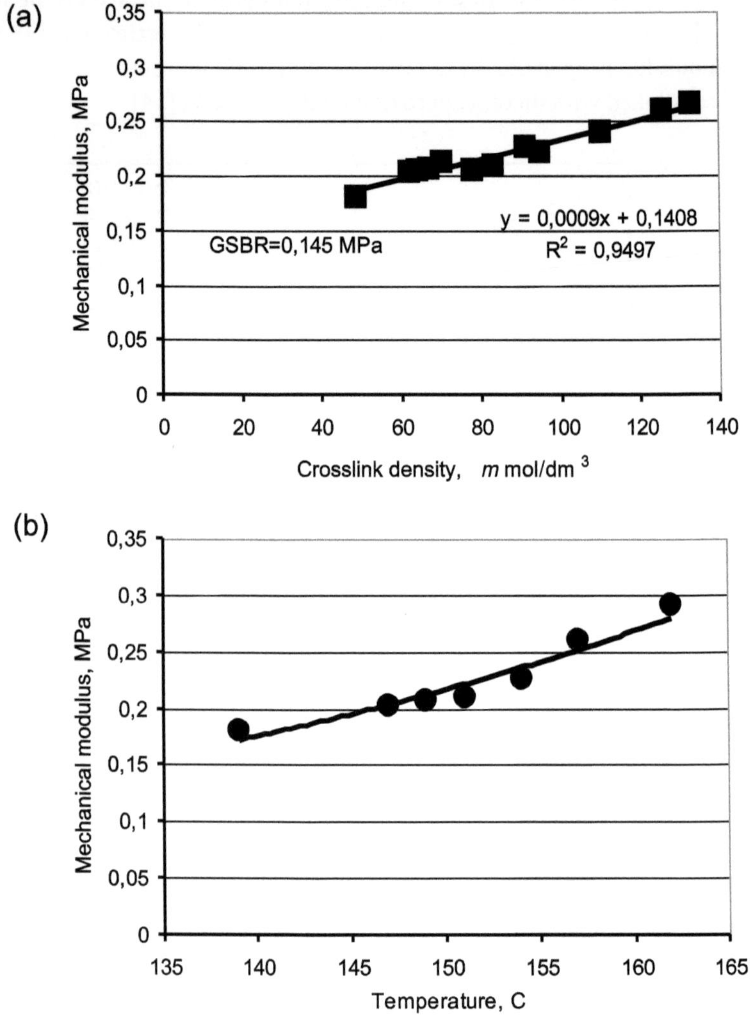

Figure 2.5 Relation between crosslink density (a) or temperature (b) and the mechanical modulus of peroxide vulcanizates of isoprene rubber.

crosslinks), which has been demonstrated for IR vulcanizates (Fig. 2.6).

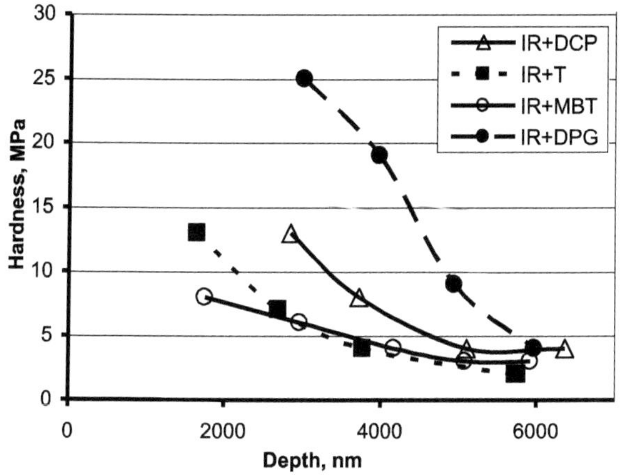

Figure 2.6 Hardness profile of isoprene rubber vulcanizates. IR/DCP: peroxide vulcanizate (data presented for comparison); IR/TMTD: rubber crosslinked with tetra(methylthiuram) disulfide; IR/S + MBT: rubber crosslinked with sulfur + mercaptobenzothiazole; IR/S + DPG: rubber crosslinked with sulfur + diphenylguanidine.

In this case the surface gradient of the crosslink density reaches ca. 80 mmol/dm^3, a value similar to the bulk crosslink density of the rubber vulcanizates studied. The superposition of temperature and structural effects results in the curve of a polynominal course.

2.3.2 *Surface Segregation in Polymer Blends*

The blooming of low-molecular-weight additives in polymer mixes concerns the components of curing systems and the products of their chemical reactions, plasticizers, and protective waxes, as well as the promoters of processing. This phenomenon can be demonstrated by the example of the peroxide vulcanizates of SBR, containing *n*-fatty acids of various lengths of a hydrocarbon backbone (Fig. 2.7) [17].

The strongest driving force for the surface migration of carboxylic acids from a rubber matrix is undoubtedly their crystallization, which takes place more easily on the surface than in the

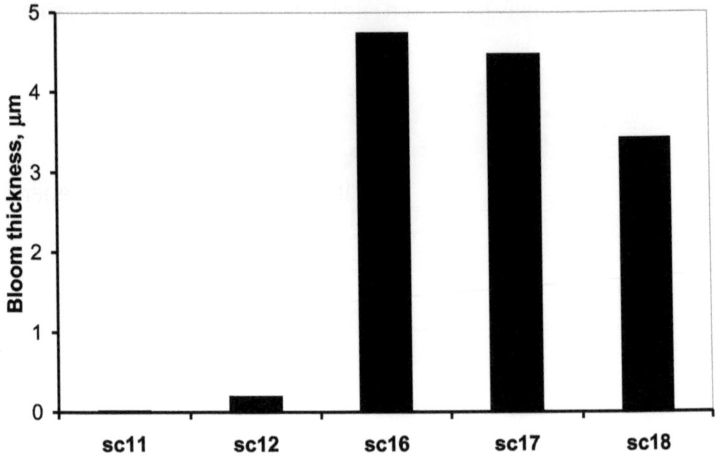

Figure 2.7 Bloom thickness of low-molecular-weight carboxylic acids present on the surface of peroxide SBR vulcanizates containing 5 phr of sc11: undecanoic acid; sc12: dodecanoic acid; sc16: hexadecanoic acid; sc17: palmitinic acid; or sc18: stearic acid.

bulk of vulcanizates. The ability of crystallization of the *n*-fatty acids studied increases with the length of their backbones but only up to 16–17 carbon atoms being present. In the case of stearic acid its surface migration becomes more difficult because of specific interactions (hydrogen bonding) themselves forming between its particles [18], which results in the decreased thickness of the bloom present on the surface of SBR vulcanizates. However, not only low-molecular-weight blooms can appear on the rubber surface; surface segregation has also been confirmed in the case of elastomer–plastomer blends [19–21].

This time most likely is the migration of the low-molecular-weight fraction of a plastomer, which, depending on its molecular weight, the difference in solubility parameters of the components, and processing conditions, manifests the ability to segregate onto the blend surface. This phenomenon has been practically applied for the modification of diene rubbers, for example, SBR, by admixing of polyolefines, such as ethylene–propylene rubber (EPM) or low-density polyethylene (LDPE), in order to improve the aging resistance of rubber goods.

An interesting system is the mixture of ethylene-propylene-diene rubber (EPDM) with LDPE. In this case an amorphous rubber matrix of a chemical structure similar to a plastomer solvates the crystalline phase of polyethylene during the blend preparation. Taking place afterward, recrystallization of LDPE leads to crystalline blooms appearing on the blend surface. Their structure depends on the molecular weight of the plastomer and its distribution, so, also in this case, the driving force for migration can be assigned to crystallization, which takes place more easily in the rubber of an amorphous matrix in comparison to the sequenced one. The crystalline phase being present on the surface makes the hardness of LDPE/EPDM blends increased significantly (Fig. 2.8).

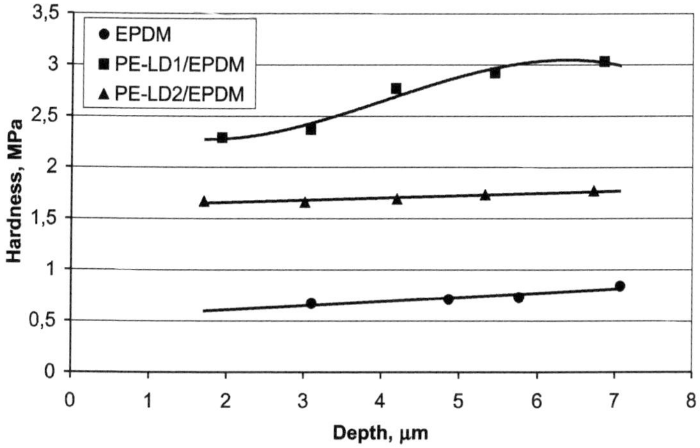

Figure 2.8 Influence of the molecular weight of low-density polyethylene (LDPE) on its surface segregation in the blends (15 phr) with ethylene-propylene-diene rubber (EPDM). LDPE 1 Mw = 15,000; LDPE 2 Mw = 35,000.

2.3.3 *Physical and Chemical Modification of the Surface Layer of Polymers*

It is known that the physical and chemical treatment of a polymer surface influence its surface energy, adhesion, tribological properties, and even thermal stability. The most popular examples are low-temperature plasma [22, 23], low-energy ion bombardment [24, 25], chlorination and sulfonation [26], or fluorination [27].

Changes to the mechanical properties and chemical composition of the surface layer of car windshield wipers made of natural rubber (NR), subjected to UV radiation treatment, are presented in Table 2.1 [28].

Table 2.1 Influence of UV radiation on the mechanical properties and chemical composition of car windshield wipers made of NR

	Vulcanization parameters						
	Time of treatment (hr)						
Properties	0	0.25	0.5	1	2	4	24
Hardness, H (MPa)	8.0	7.5	6.5	6.0	4.0	3.0	4.0
Mechanical modulus, E (MPa)	25.5	28.0	28.5	28.0	17.0	16.0	22.0
H/E	0.31	0.27	0.30	0.21	0.24	0.19	0.18
Hydrogen content (at.%)	66	62	53	50	42	40	25

UV radiation is commonly believed to be a negative factor, responsible for the aging of polymers. However, the results obtained demonstrated that applied in a moderate dose it can lead to the positive modification of rubber. Due to the treatment the hardness and mechanical modulus of its surface layer deteriorate, which is the result of material graphitization taking place. However, this fact facilitates the improvement of wear and aging resistance, as well as the significant lowering of vulcanizate friction.

Also application of ion bombardment for modification of polymer materials, in comparison to metals or ceramics, has a relatively short history [28]. One can find only occasional contributions devoted to the structural and surface engineering aspects of polymer materials. However, contrary to general concerns on possible material degradation, the proper selection of ion mass, chemical reactivity toward targets, ion fluence, and beam energy makes it possible not only to improve the durability of polymers but also to produce materials of interesting characteristics. The application of ion bombardment to polymer materials results in significant changes to the composition and structure of their surface layer. The modification releases hydrogen, which often saturates at an atomic ratio C:H = 1 [29]. Depending on the macromolecular structure of the polymer target it results either in crosslinking or in degradation of the material [30], which is reflected by the changes in hardness (Fig. 2.9).

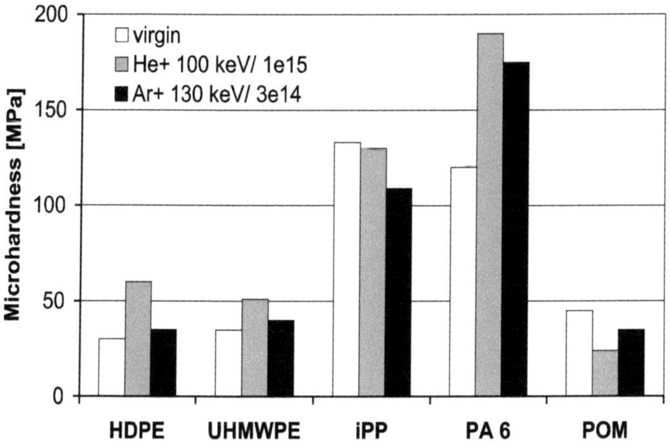

Figure 2.9 Influence of ion bombardment on hardness of some polymers studied. HDPE, high-density polyethylene; UHMWPE, ultra-high molecular weight polyethylene; iPP, isotactic polypropylene; PA6, polyamide 6; POM, polyoxymethylene.

The effect of modification, apart from the kind of ions and the fluence applied, depends on the molecular weight and supermolecular structure of polymers [31]. Ion bombardment results in the oxidation of the surface layer [32], which together with the modification of the surface geometry, influences the wettability of polymers.

The effect of interaction between an ion beam and rubber macromolecules seems to be mainly energetic [33], resulting in chain scission and the arising of macroradicals, as well as ionization-manifesting itself by the release of hydrogen. Depending on the molecular structure of rubber, its macromolecules exhibit a tendency either to degradation (natural and chloroprene rubber) or to crosslinking (styrene–butadiene and acrylonitrile–butadiene rubbers). Apart from hydrogen release resulting in graphitization, free radicals, produced in the surface layer of rubber vulcanizates by ion bombardment, react readily with atmospheric oxygen, modifying chemically the materials. The effect is durable and the

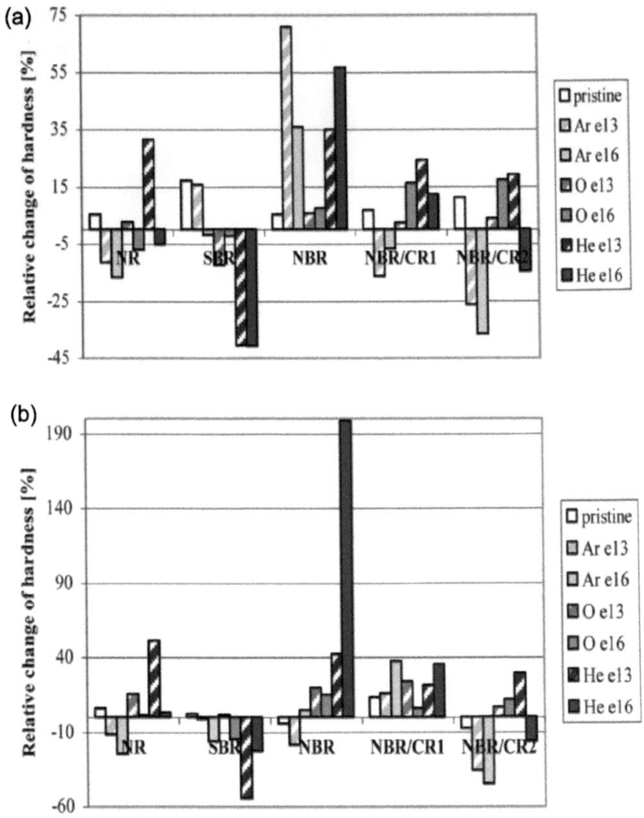

Figure 2.10 The influence of 70 keV ion bombardment on the changes to the nanoindentation hardness of rubber vulcanizates subjected to thermal (a) and ozone (b) aging. NR, natural rubber; SBR, styrene–butadiene rubber; NBR, butadiene–acrylonitrile rubber; NBR/CR, blend of butadiene–acrylonitrile and chloroprene rubbers.

extent of modification limited to the range of the ion beam, as demonstrated by Rutherford backscattering spectrometry (RBS) oxidation profiles. It looks like oxygen gradually saturates the surface layer of the rubber vulcanizate, preventing the bulk of the material from deeper penetration of oxygen. This observation creates a possibility for application of ion bombardment to protect rubber vulcanizates from aging (Fig. 2.10) or swelling in fuels (Fig. 2.11), which will be also discussed later.

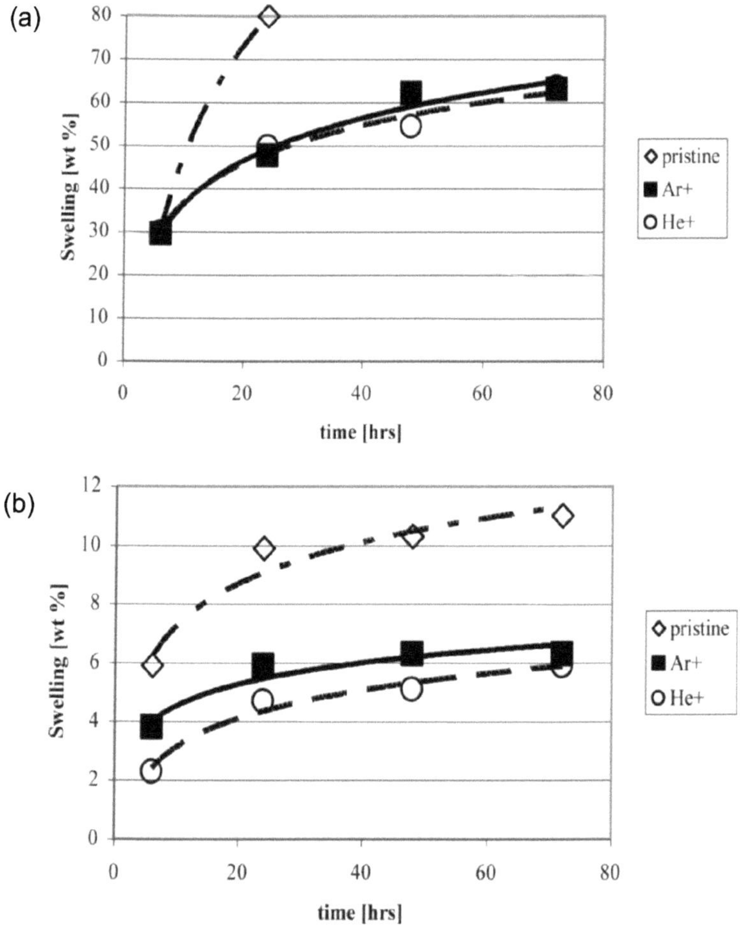

Figure 2.11 The influence of ion bombardment (1×10^{15} 1/cm^2/70 keV) on the changes to the nanoindentation hardness of rubber vulcanizates subjected to swelling in the standard fuel (ISO 1817). (a) Unpolar material, NR; (b) polar material, NBR/CR. NR, natural rubber; NBR/CR, blend of butadiene–acrylonitrile and chloroprene rubbers.

The ability of ion implantation to tailor hardness and wettability allows one to reduce the friction and wear of the materials. Ion beam bombardment has been found to be effective, in terms of reduction friction, toward polymers and elastomers for which conventional chemical treatments (halogenation or sulfonation) fail (Fig. 2.12).

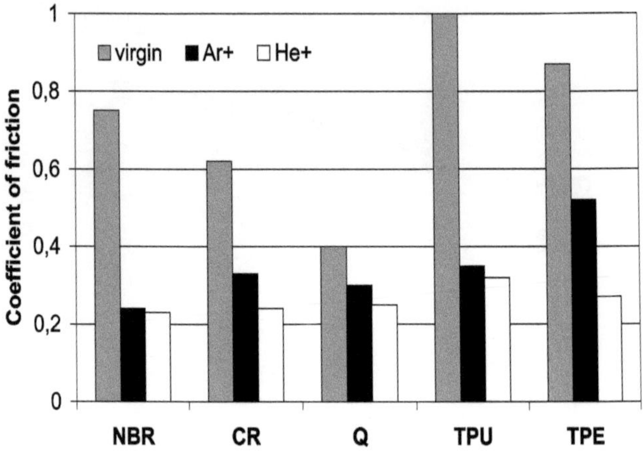

Figure 2.12 The influence of ion bombardment on the friction coefficient of some elastomers studied. NBR, butadiene–acrylonitrile rubber; CR, chloroprene rubber; Q, silicone rubber; TPU, thermoplastic polyurethane; TPE, thermoplastic elastomer (product of dynamic covulcanization).

The most popular method of chemical modification is chlorination because of its low cost and simple technology. The process most often used to be carried out in liquid media, monitoring the chlorine ion concentration in solution and the time of treatment. Strong oxidation of the surface layer, usually accompanying too intensive chlorination, results in significant structural changes, which manifest themselves by changes to mechanical characteristics of the material. The possibility for application of nanoindentation to optimize the time of treatment from the point of view of rubber hardness has been demonstrated in the example of car windshield wipers (Fig. 2.13) [34].

In other studies, nanoindentation served the selection of a material and the parameters of its thermal treatment for making the runners of annular textile machines (Table 2.2).

On the basis of hardness and mechanical modulus measurements it was possible to select the most suitable kind of polyamide material in order to make a choice of modification medium (oil or water) and to optimize the temperature and time of thermal annealing [35, 36]. The changes to the mechanical characteristics of the surface layer of the polymer material were correlated with the changes to its

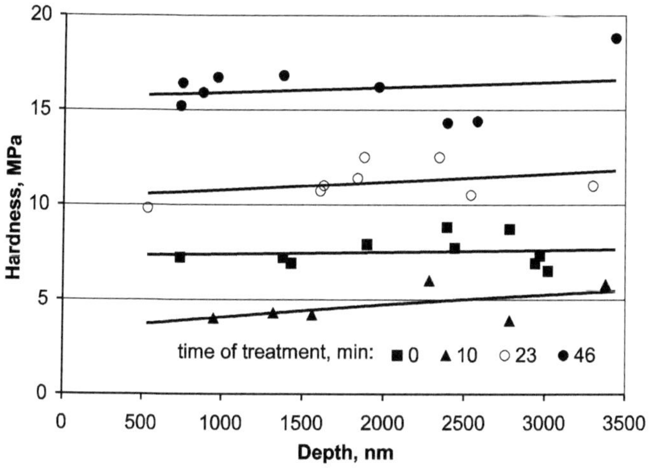

Figure 2.13 The influence of chlorination on hardness of car windshield wipers made of NR.

Table 2.2 The influence of heat treatment on the hardness and mechanical properties of polyamide runners of textile machines

Material	Hardness (MPa)	Relaxation (%)	Elastic recovery (%)
PA 66 virgin	57.4	20.3	21.3
PA 66 after oil annealing	87.4	15.1	15.1
PA 66 after water annealing	89.3	10.3	10.5
PA 6 virgin	113	11.7	11.1
PA 6 after oil annealing	127	11.2	13.7
PA 6 after water annealing	50.1	7.6	8.0

structure and application parameters, measured by the values of the coefficient of friction and the abrasion resistance against polished stainless steel.

2.3.4 *Photopolymerization of Polyester Resin*

Due to the possibility of making measurements also in the range of elastic deformations, nanoindentation, opposite to the classical hardness measurement, enables studying of the mechanical properties of thin films.

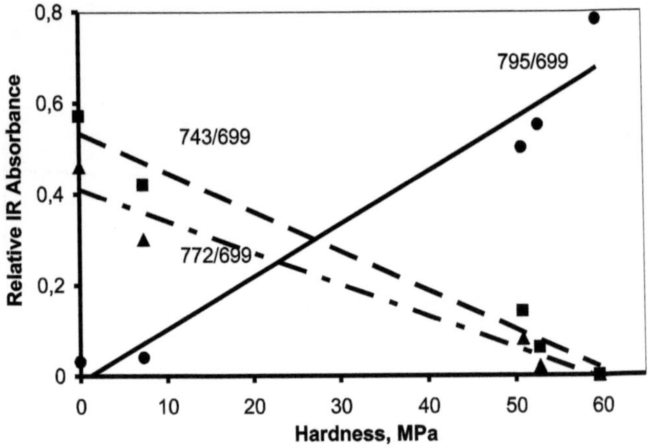

Figure 2.14 The application of nanoindentation for the estimation of the curing degree of polyester resin coating.

Figure 2.14 demonstrates that the changes to coating hardness resulted from photopolymerization can be attributed to the conversion of polyester resin-measured by the decrease of double bonds (abs. at 772 cm^{-1}) and with the polarization or graphitization of the surface layer-adequate either to the degree of its oxidation (abs. at 795 cm^{-1}) or to the decrease of hydrogen content (abs. at 743 cm^{-1}) [37].

Experiences gained during studies on the photopolymerization of thin resin films have been utilized in the further investigations of polymer composites for dental restoration. The application of nanoindentation makes it possible to monitor the photopolymerization degree of dimethacrylate resin, constituting a base of the composite, as well as the influence of fillers on the course of curing, deciding both on the exploitation properties of dental filling [38].

2.3.5 *Aging of Polymer Materials*

Up to now the resistance of polymers to thermal or ozone aging used to be estimated from changes to their mechanical proprieties, determined by conventional tensile strength measurements (PN-ISO 11346), while chemical and physical changes to the materials,

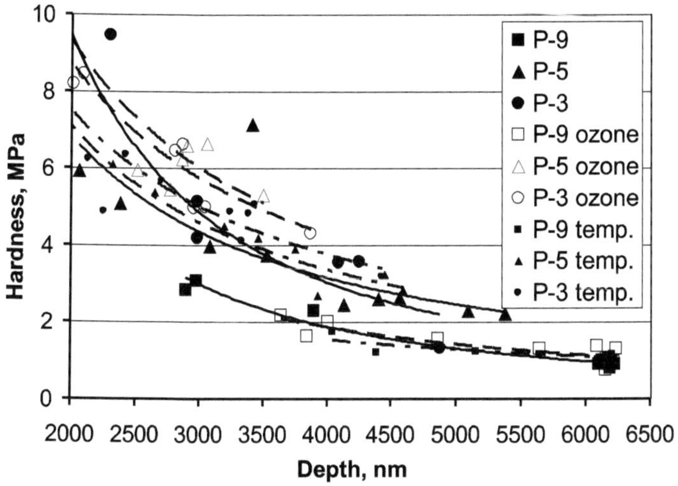

Figure 2.15 The influence of aging on the surface profile of the IIR composites' hardness.

being the result of elevated temperature or the action of ozone, are initiated and usually limited to their surface layer. The application of nanoindentation makes it possible to detect the beginning of material degradation, which starts from the surface and gradually propagates deeper inside a polymer. It is very important because it creates the opportunity to make the right decision on the further exploitation of responsible parts made of polymers at the right time. It has been illustrated in the example of thin composite films made of buthyl rubber (IIR) filled with montmorillonite (MMT), subjected to thermal (temp.) or ozone (ozone) aging (Fig. 2.15) [5].

Aging processes result in a change to the hardness of polymer material. From Fig. 2.15 it follows that most resistant to aging is composite P-9, for which the hardness profiles after thermal or ozone treatment remain practically unchanged comparing to the virgin material. Composite P-5 is resistant only to thermal aging, whereas composite P-3 does not exhibit any resistivity to aging, no matter whether of thermal or ozone origin. In the light of conducted research it has been shown that morphology, shaped already at the stage of processing, is the most important factor, deciding the susceptibility of the materials studied to aging [39].

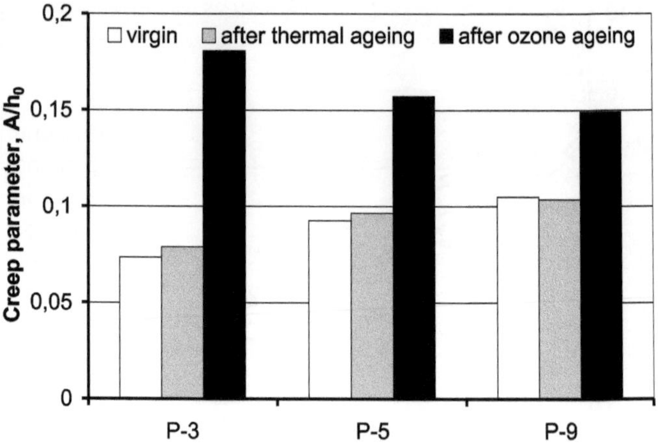

Figure 2.16 The influence of aging on creep behavior of the IIR/MMT composites studied.

The comparison of changes to the profile of material hardness due to aging gives only the qualitative information on the aging resistance of polymer materials. The quantitative analysis can be obtained calculating the so-called creep coefficient, proposed recently in the subject literature [11, 12]. Figure 2.16 presents the results calculated for the thin polymer films studied.

The creep results presented do not confirm the hardness results calculated from nanoindentation data. Nevertheless the smallest changes of the undimensional creep parameter (A/h_0) have been also noted in the case of composite P-9.

2.3.6 *Surface Gradient of Rubber Vulcanizate Crosslinks*

The exploitation of fuel engines at more and more higher temperature, being the consequence of the race for the highest efficiency, together with the application of alternative fuels containing bio-components, leads to the situation that rubber products faced more and more harder requirements from industry. Therefore, industry is looking for new materials or modifies already existing materials, making them being able to work in a biofuel environment at elevated temperature.

Table 2.3 The influence of exploitation (biodiesel, 160°C) on the hardness profile of fuel hoses made of various acrylic rubbers (VAMAC)

	Properties			
	Virgin sample		Sample after exploitation	
Vulcanizate	Force (mN)	Hardness (MPa)	Force (mN)	Hardness (MPa)
	0.28	1.55	0.30	1.64
	0.29	1.58	0.29	1.59
	0.27	1.48	0.26	1.41
VAMAC 1	0.27	1.48	0.31	1.68
	0.28	1.58	0.31	1.70
	0.28	1.53	0.28	1.58
	0.27	1.51	0.28	1.53
Median	0.28	1.53	0.29	1.59
	0.33	1.84	0.45	2.46
	0.34	1.88	0.42	2.32
	0.32	1.77	0.44	2.43
VAMAC 2	0.32	1.79	0.43	2.37
	0.34	1.83	0.45	2.46
	0.35	1.95	0.46	2.55
	0.34	1.87	0.45	2.51
Median	0.33	1.85	0.44	2.44
	0.33	1.83	0.44	2.43
	0.31	1.76	0.47	2.59
	0.33	1.84	0.46	2.54
VAMAC 3	0.33	1.78	0.48	2.66
	0.32	1.78	0.48	2.57
	0.34	1.84	0.44	2.42
	0.34	1.86	0.45	2.49
Median	0.33	1.81	0.46	2.53

In Table 2.3 the hardness profile of three commercial elastomers based on acrylic rubber (VAMAC), after six-month service in pumping biodiesel at a temperature of 160°C, have been compared.

From the results obtained it follows univocally that most resistant to the action of hot fuel is the material VAMAC 1. Its hardness, among the samples studied, increases to the least extent after the exploitation period, which can be attributed to the lowest degree of polymer degradation. In this case the degradation process manifests itself by the increase of hardness, being the result of the higher degree of crosslinking produced by the oxidation of rubber and the extraction of low-molecular-weight components from rubber mixes. Similarly to aging, also in the case of swelling, kinetics of the process is determined from volume measurements, whereas both phenomena originate in the surface layer of polymers. Changes taking place in the surface layer, being very often responsible for the failure of parts made of polymer materials, are not able to be detected applying conventional volume methods.

2.4 Summary and Conclusions

Micromechanical characteristics of the surface layer of polymer materials, obtained by nanoindentation, make it possible to monitor/control the changes taking place in their top surface, for example, being the result of aging, chemical (halogenation, sulfonation) and physical (ionizing radiation, ion bombardment) modification, or swelling in fuels and other media. Due to the surface-sensitive character of the changes they are not able to be detected applying the traditional volume approach. The progress of the above-mentioned processes or phenomena can be estimated from the changes to hardness of the material, determined by nanoindentation (ΔH). Another source of information can be the creep characteristic of the surface layer of polymer materials, changing due to their physical or chemical treatment or modification, being the result of aging or swelling. The degree of material modification or their aging resistance can be determined quantitatively on the basis of the changes to the so-called undimensional creep parameter (A/h_0).

References

1. Burakowski, T., and Wierzchoń, T. (1971). *Engineering of Metal Surface* (WNT, Warsaw) (in Polish).
2. Bieliński, D.M. (2004). On importance of the surface layer for exploitation of polymer materials, *Kautsch. Gummi Kunstst.*, **57**, 13–21.
3. Bieliński, D.M., and Ślusarski, L. (2000). Elastomer materials, in *Tribology and Tribotechnics*, Szczerek, M., and Wiśniewski, M., eds. (ITeE Press, Radom), 405–413 (in Polish).

4. Ślusarski, L., Bieliński, D.M., Affrossman, S., and Pethrick, R.A. (1998). Surface modification of elastomers toimprove their tribological properties, *Kautsch. Gummi Kunstst.*, **51**, 429–438.
5. Bieliński, D.M. (2009). Application of microindentation to study ageing of rubber, *Kautsch. Gummi Kunstst.*, **62**, 28–32.
6. www.micromaterials.co.uk
7. Bieliński, D.M., and Ślusarski, L. (2010). Positron annihilation lifetime spectroscopy (PALS) and nanoindentation (NI), in *Rubber Nanocomposites: Preparation, Properties and Applications*, Sabu, T., and Ranimol, S., eds. (T. J. Wiley and Sons, Singapore), 595–630.
8. Oliver, W.C., and Pharr, G.M. (1992). An improved technique for determining hardness and elastic modulus using load and displacement sensing indentation experiments, *J. Mater. Res.*, **7**, 1564–1583.
9. Beake, B.D., Chen, S., Hull, J.B., and Gao, F. (2002). Nanoindentation behaviour of clay/poly(ethylene oxide) nanocomposites, *J. NanoSci. Nanotech.*, **2**, 73–79.
10. Beake, B.D., Zheng, S., and Alexander, M.R. (2002). Nanoindentation testing of plasma-polymerized hexane films, *J. Mater. Sci.*, **37**, 3821–3826.
11. Beake, B.D., Goodes, S.R., Smith, J.F., and Gao, F. (2004). Nanoscale repetitive impact testing of polymer films, *J. Mater. Res.*, **19**, 237–247.
12. Beake, B.D. (2006). Modelling indentation creep of polymers: a phenomenological approach, *J Phys D: Appl Phys.*, **39**, 4478–4485.
13. Bieliński, D.M. (2009). *Tribology of Elastomers and Rubber from Material Engineering Point of View* (ITeE-PIB Press, Radom) (in Polish).
14. Bieliński, D.M., Kajzer, M., Ślusarski, L., and Kaczmarek, Ł. (2004). Gradient structure of rubber vulcanizatess, *Polimery*, **49**, 298–304 (in Polish).
15. Bieliński, D.M. (2004). Gradient structure of polymer materials, *Curr. Trends Polym. Sci.*, **9**, 33–45.
16. Boochathum, P., and Prajudtake, W. (2001). Vulcanization of cis- and trans-polyisoprene and their blends: cure characteristics and crosslink distribution, *Eur. Polym. J.*, **37**, 417–427.
17. Bieliński, D.M., Głąb, P., and Ślusarski, L. (2005). Surface segregation of carboxylic acids in styrene-butadiene rubber: effect of the bloom on friction, *Compos. Interfaces*, **12**, 445–457.

18. Bieliński, D.M., Ślusarski, L., Boiteux, G., and Parasiewicz, W. (2000). Migration of low molecular weight substances in elastomer matrix. Structure and properties of top layers formed, *Proc. 38th World Polymer Congress "IUPAC Macro,"* 1175–1176.
19. Janczak, K.J., Janczak, T., and Ślusarski, L. (1989). Friction and wear of composite polymer material, *Wear*, **130**, 93–101.
20. Bieliński, D.M., Ślusarski, L., Włochowicz, A., and Douillard, A. (1997). Unusual behaviour of polyethylene in ethylene-propylene-diene matrix, *Compos. Interfaces*, **5**, 155–178.
21. Bieliński, D.M., and Kaczmarek, Ł. (2006). Surface segregation of polyethylene in LDPE/EPDM blends. Aspects of components' structure, *J. Appl. Polym. Sci.*, **100**, 625–633.
22. Lawson, D.F. (1987). Corona discharge activation and reconstruction of elastomer surfaces, *Rubber Chem. Technol.*, **60**, 102–110.
23. Grythe, K.F., and Hansen, F.K. (2006). Surface modification of EPDM rubber by plasma treatment, *Langmuir*, **22**, 6109–6124.
24. Turos, A., Jagielski, J., Piątkowska, A., Bieliński, D.M., Ślusarski, L., and Madi, N.K. (2003). Ion beam modification of surface properties of polyethylene, *Vacuum*, **70**, 201–206.
25. Ostaszewska, U., Pieczyńska, D., Bieliński, D.M., and Jagielski, J. (2012). Modification of polymers by ion bombardment. Part II: modification of functional properties of rubber, *Polimery*, **57**, 40–50 (in Polish).
26. Bieliński, D.M., Ślusarski, L., Hartshorne, M., Affrossman, S., and Pethrick, R.A. (1995). Influence of chemical modification on tribological properties of elastomers, *J. Appl. Polym. Sci.*, **56**, 853–867.
27. Anyszka, R., Bieliński, D.M., Mężyński, J., Grams, J., Rehwinkel, C., and Möller, B. (2010). Influence of surface fluorination of rubber on its tribological properties, *Tribologia*, **232**, 13–21 (in Polish).
28. Kondyurin, A., and Bilek, M. (2008). *Iop Beam Treatment of Polymers. Application Aspects from Medicine to Space* (Elsevier, Oxford).
29. Bieliński, D.M., Lipiński, P., Ślusarski, L., Grams, J., Paryjczak, T., Jagielski, J., Turos, A., and Madi, N.K. (2004). Surface layer modification of ion bombarded HDPE, *Surf. Sci.*, **564**, 179–186.
30. Dong, H., and Bull, T. (1999). State-of-the-art overview: ion beam surface modification of polymers towards improving tribological properties, *Surf. Coat. Technol.*, **111**, 29–40.
31. Bieliński, D.M., Tranchida, D., Lipiński, P., Jagielski, J., and Turos, A. (2007). Ion bombardment ofpolyethylene: influence of polymer structure, *Vaccum*, **81**, 1256–1260.

32. Jagielski, J., Turos, A., Abdul-Kader, A.M., Piątkowska, A., and Bieliński, D.M. (2007). Ion beam modified polymers for biomedical applications, *Nucl. Instrum. Methods*, **B261**, 690–693.
33. Jagielski, J., Grambole, D., Jóźwik, I., Bieliński, D.M., Ostaszewska, U., and Pieczyńska, D. (2011). Hydrogen loss from elastomers subjected to ion irradiation, *Mater. Chem. Phys.*, **127**, 342–346.
34. Bieliński, D.M., Głąb, P., and Ślusarski L. (2006). New approach to study tribological properties of polymer materials. Case of windshield wipers, *J. Achiev. Manuf. Eng.*, **15**, 71–78.
35. Bieliński, D.M., Wójcicki, R., and Ambroziak, H. (2003). Studies of polymer runners for annular textile machines. Part I: physical and chemical characteristics of the materials, *Exploit. Probl.*, **51**, 67–82 (in Polish).
36. Bieliński, D.M., Wójcicki, R., and Ambroziak, H. (2004). Studies of polymer runners for annular textile machines. Part II: physical modification and thermal treatment, *Exploit. Probl.*, **52**, 29–40 (in Polish).
37. Bieliński, D.M., Głąb, P., and Ślusarski, L. (2001). Application of FT-IR technique and nanoindentation measurements for studying progress in photocuring of polyester resin, *Polimery*, **7/8**, 494–500 (in Polish).
38. Kleczewska, J., Sokołowski, J., Bieliński, D.M., and Klimek, L. (2007). Influence of surface layer morphology on abrasion of polymer dental composites, *Mater. Eng.*, **28**, 930–934 (in Polish).
39. Bieliński, D.M. (2009). Elastomer nanocomposites: structure, properties and applications. Current state and perspectives of progress, in *Studies of the Polish Committee of Technical Sciences of the Polish Academy of Arts & Sciences 3* (Polish Academy of Arts and Science Press, Cracow), 165–187 (in Polish).

Chapter 3

Characterization of Light-Cured Dental Composites

Joanna Kleczewska,[a] Dariusz M. Bieliński,[a] Narayanaswami Ranganathan,[b] and Jerzy Sokolowski[c]

[a]*Institute of Polymer & Dye Technology, Lodz University of Technology, Stefanowskiego 12/16, 90-924 Lodz, Poland*
[b]*Laboratoire de Mécanique et de Rhéologie (LMR) EA2640, Université François Rabelais de Tours (UFRT), Polytech 'Tours, 7 avenue Marcel Dassaul, Tours, 37200, France*
[c]*Department of General Dentistry, Medical University of Lodz, Pomorska 251, 92-213 Lodz, Poland*
joanna.kleczewska@p.lodz.pl

3.1 Introduction

The problem of reconstruction of damaged parts of the human body, including teeth, is one of the most important issues in medicine. Today, research on the possibility of tooth replacement by artificial materials is one of the fastest-growing branches of dentistry.

The idea of modern restorative dentistry is to keep pace with nature—create a "chameleon," which would be similar to natural tooth tissues—both visually and physically. Over the past several years, the dental materials used as a filling for cavities in teeth have

Materials Characterization: Modern Methods and Applications
Edited by Narayanaswami Ranganathan

ISBN 978-981-4613-06-4 (Hardcover), 978-981-4613-07-1 (eBook)
www.panstanford.com

evolved into a truly satisfactory form. They allow us to reproduce damaged tooth tissue precisely, but there are still several important problems that should be solved finally. The most important of them seem to be [1]:

- Reducing the "polymerization" shrinkage, resulting in improved marginal adaptation of the fillings.
- Improving the mechanical properties of fillings, resulting in longer lifetime of composites.
- Providing bactericidal function to composites—this will prevent accumulation of dental plaque and secondary caries formation.

These three issues are the most serious challenges for dental composite engineering. This is the reason why the development of dentistry in recent years has caused the materials on the market today to be very different from those a few decades ago.

3.1.1 *Chemistry of Dental Composites (Yesterday and Today)*

3.1.1.1 Matrix

One of the first (1941) aesthetic polymer dental filling was based on methacrylate monomers—methyl polymethacrylate dissolved in methyl methacrylate. As a radical polymerization initiator system, tertiary amines were used in combination with benzoyl peroxide. Thanks to this combination of activator (amine) and initiator (peroxide), the crosslinking reaction could be carried out at the temperature that occurs in the mouth and finished within a few minutes. However, these materials exhibited significant polymerization shrinkage (20%–25%), susceptibility to discoloration, reduced stiffness, high thermal expansion, and poor adhesion to the tooth surface.

In 1957, Bowen patented a new type of resin, which combined the positive features of methacrylates (fast polymerization) with the advantages of epoxy resins (better adhesion to the tooth surface and higher mechanical strength). The new resin, known as bis-GMA (2,2-bis-[4,4-(2′-hydroxy-3′-methacryloxypropoxy)phenyl]propane) (Fig. 3.1), produced in the reaction of

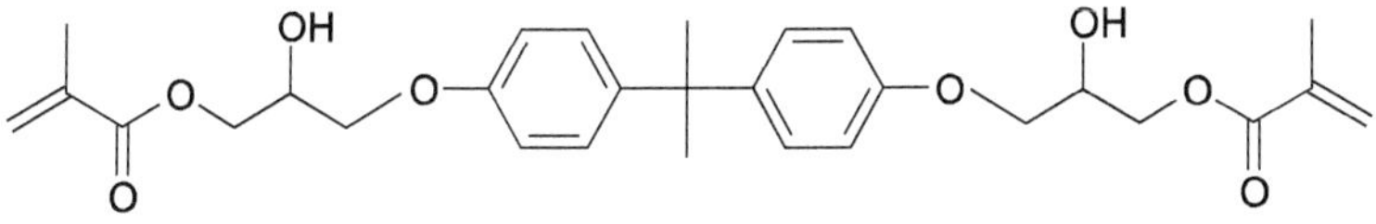

Figure 3.1 Bis-GMA resin, 2,2-bis-[4,4-(2′-hydroxy-3′-methacryloxypropoxy)phenyl]propane.

bisphenol-A with glycidyl methacrylate, made an outstanding career and to this time is an essential component of the dental composite matrix.

The advantage of bis-GMA resin is a rigid aromatic structure and low water sorption. The three-dimensional spatial structure of bis-GMA, formed during polymerization, is characterized by high entanglement of macromolecules chains, linked by secondary van der Waals bonds. Because of this, the glass transition temperature and melting point of bis-GMA are higher. Additionally, due to higher molecular weight and the presence of an aromatic ring, the application of bis-GMA has reduced the value of shrinkage to about 5% and increased the mechanical strength of the matrix [2, 3].

Bowen's invention has contributed to the rapid replacement of dental cements and acrylic resins by the dimethacrylate composites. However, due to the fact that the increase in molecular weight of the main component of the matrix resulted in a significant increase in its viscosity, it was necessary to find an appropriate reactive diluent—a substance that reduces viscosity of bis-GMA resin, additionally participates actively in the polymerization reaction, and facilitates the introduction of mineral filler.

Today, triethylene glycol dimethacrylate (TEGDMA) is the most commonly used diluent. It changes the viscosity of the system and improves the flexibility of the final composite [4, 5]. The TEGDMA resin structure is similar to bis-GMA (Fig. 3.2), that is, it has reactive double bonds in the methacrylate groups located at both ends of the molecule.

However, the whole molecule is shorter, and it does not contain aromatic rings—has lower rigidity than bis-GMA, resulting in viscosity several orders lower (Table 3.1) [6].

The smaller molecular weight of TEGDMA is responsible for significant shrinkage (approximately 13%) during polymerization

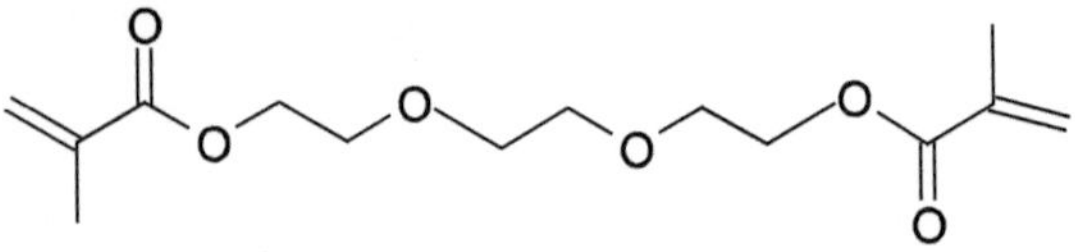

Figure 3.2 TEGDMA resin, triethylene glycol dimethacrylate.

Table 3.1 The viscosity of different bis-GMA/TEGDMA mixtures [6]

Bis-GMA/TEGDMA (wt.%/wt.%)	Viscosity (cP)
100/0	1.2×10^6
76.2/23.8	5240
50/50	282
33.3/66.7	245
0/100	15

[3]. If we look at the problem only through the prism of shrinkage, we should conclude that the addition of TEGDMA to the matrix is disadvantageous, but actually, it is worth of note that it is not possible to introduce the filler into the matrix without reducing the viscosity of bis-GMA resin.

In 1974, Foster and Walker patented a new type of bifunctional urethane dimethacrylate (UDMA) resin (Fig. 3.3) [7]. This type of oligomer is formed by the reaction of oligo(ethylene) glycol monomethacrylates containing from one to four oxyethylene units (e.g., HEMA, ethylene glycol monomethacrylate) with different diisocyanates (such as HMDI, hexamethylene diisocyanate) [8]. UDMA resins are characterized by lower water absorption and smaller viscosity in comparison to bis-GMA, so the necessity of dilution of UDMA monomers is redundant. Unfortunately, UDMA resins are also more hard and brittle, and their shrinkage is quite large (5%–9%). In addition, this type of resin is sometimes less resistant to light and may be prone to discoloration [9, 10].

Despite time having passed and the progress of science, these imperfect oligomers—bisGMA, UDMA, and TEGDMA—are still essential components of most commercial dental composites. However, the possibility of modifying the matrix resin in the context

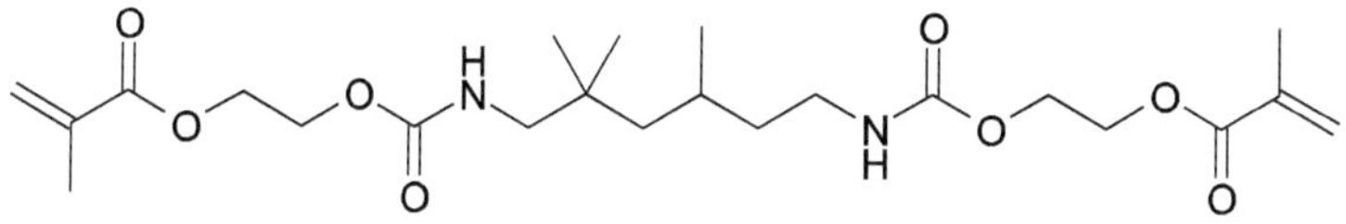

Figure 3.3 UDMA resin, 1,6-bis(methacryloxy-2-ethoxycarbonyl amino)-2,2,4-trimethyl hexane.

of polymerization shrinkage reduction is still one of the most exploited issues.

Recent studies have shown that silorane-based materials exhibit lower shrinkage (1%), accompanied by the smaller shrinkage stress, ??than most commercially available dental composites [11–14]. In terms of chemical structure, siloranes are completely different from all previously proposed materials (Fig. 3.4).

It entails far-reaching consequences: the new photoinitiation system (besides camphorquinone, salt iodonium, and electron donor also), cationic polymerization, and epoxy-based functional groups. These differences make siloranes totally incompatible with all existing dentin-composite binding systems. Thereby, application of them successfully in dentistry practice requires much knowledge and individual treatment. Silorane materials found their real place

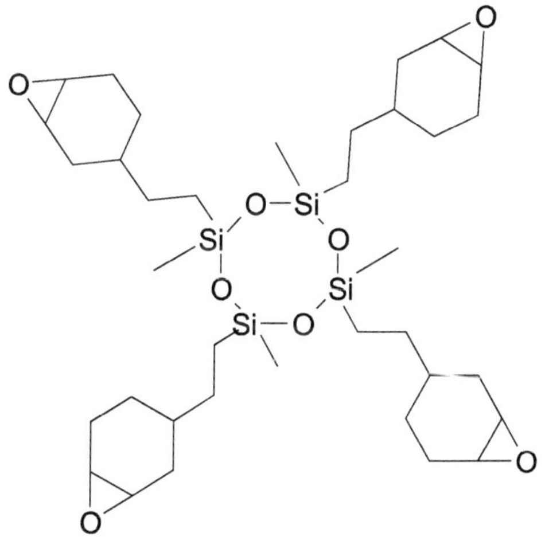

Figure 3.4 Silorane.

on the commercial market next to the dimethacrylate ones and have been offered for sale from autumn 2007 [14].

3.1.1.2 Fillers

An inorganic phase is—next to the organic phase—an integral constituent of composites. In dental materials, the role of the inorganic phase is that of a filler, usually added in an amount of 75–90 wt.% (about 35–60 vol.%) [15]. The usage of such a huge quantity of filler is primarily (but not exclusively) the result of efforts to minimize the volumetric shrinkage of the resin. It is well known that a high degree of filler content and good dispersion of the inorganic phase in the resin matrix decrease the size of the volumetric shrinkage. In the case of commercial composites containing 60–70 vol.% filler, volumetric shrinkage is approximately 2%, while the composites filled to a lesser extent (47 vol.%) can change their volume during the polymerization by up to 3.5% [16].

The inorganic phase in commercial dental composites is usually a powder of different-sized and different-shaped particles, inert with respect to the copolymers (for this reason it is generally subjected to modification—silanization). Generally, the filler phase in dental composites constitutes different silicon compounds: dioxide (amorphous silica, quartz), borosilicate, Li–Sr–Ba–Al glass, and oxides (zirconium, aluminum). Due to the different structures and chemical nature, each of these compounds will play a slightly different function in the restorations. Probably for this reason, generally more than one type of filler is used in commercial composites.

Quartz is the hardest filler. Its hardness is greater than most glasses, and therefore, its presence in the composite results in increased resistance to abrasion. In addition, due to the appropriate refractive index, quartz enables one to obtain highly aesthetic fillings. Most of the glass fillers behave similarly. Their addition results in the so-called chameleon effect—the fillings conformed to the shade of surrounding tissues.

The introduction of elements of high atomic weight (barium, zirconium, strontium, tantalum) in dental composites makes it possible to obtain a good contrast in radiographs, which facilitates the correct diagnosis of fillings [17–19]. However, studies have shown that composites containing Ba have a tendency to barium leaching, which could deteriorate the mechanical strength of the fillings and increase the degree of abrasive wear [20].

The research on improving the inorganic phase in dental composites is currently focused on fibers: single-crystal fibers (whiskers) [21–24] or fine glass fibers [25, 26]. A number of papers that appeared in the last decade indicate that the fibers are still an interesting alternative to round-shaped particles. The fiber form appears to be more beneficial than the fine form because it provides better energy absorption by the filler. It reduces internal stresses in the material and usually results in increasing mechanical strength of the composites and decreasing susceptibility to abrasion [22, 25].

Specially crafted whisker fillers—the single crystals of an elongated shape and small size (0.1–1 μm)—are characterized by particularly high strengthening potential. The application of whiskers leads to flexural strength, even twofold higher than in commercial composites. The other mechanical properties, like resistance to scratch, abrasion, or the formation and propagation of microcracks, are also improved [21].

Unfortunately, the nonaesthetic appearance and small (and rather negative) impact on the value of shrinkage cause that fillings containing fibrous fillers not find a place on the commercial market [26].

3.1.2 *Photopolymerization: Gradient of Morphology and Properties*

Properties of dental materials can be developed by modifying the chemical composition and morphology of composites, in particular their surface layer [27]. Light-cured dental composites are multiphase and multicomponent systems. The nonuniform degree of matrix crosslinking from the surface into the bulk of material [28], which is the result of the use of the irradiation technique for curing, together with poor thermal conductivity of the polymer matrix, is the reason of the gradient of the surface composition and morphology (e.g., a surface gradient of mechanical properties, which, in consequence, leads to different resistance of materials to abrasive wear) [29].

The rate of polymerization initiation (R_i) depends on the intensity of light absorbed (I_a) [30], according to Eq. 3.1:

$$R_i = 2\o I_a \tag{3.1}$$

where I_a is the number of photons expressed in moles (1 mol of photons = einstein) absorbed per unit voume and time (per liter and second) and ø is the initiation quantum efficiency.

The intensity of light absorbed (I_a) depends on the intensity of incident light (I_0) and the absorbance of the photoinitiator (A), according to Eq. 3.2:

$$I_a = \frac{I_0(1 - 10^{-A})}{d} \tag{3.2}$$

where d is the thickness of the irradiated layer (depth of light penetration).

The absorbance (A) is proportional to the initiator concentration (In) and the layer thickness (d) (Eq. 3.3):

$$A = \varepsilon d[\mathrm{In}] \tag{3.3}$$

where ε is the molar absorption coefficient, expressed in $dm^3 \cdot mol^{-1} \cdot cm^{-1}$.

Equations 3.1–3.3 show that R_i is proportional directly to I_0 but not to the concentration of the initiator. Increasing the concentration of the initiator (and thus the absorbance) will cause more and more light to be absorbed in the upper layers of the irradiated samples, and less and less light will reach the deeper layers. It may leads to a nonuniform degree of conversion (crosslinking) of the material, the formation of internal stresses, as well as deterioration of adhesion of the polymerized coating and deterioration of its mechanical properties. This situation is illustrated in Fig. 3.5, which presents the ratio of absorbed light intensity to the total incident light intensity (I_a/I_0) as a function of absorbance. Two conclusions may be drawn from the this graph: (1) The unequal absorption occurs already at relatively small values of absorbance and increases with A and (2) the absorption in the deeper parts of the samples first increases and then decreases with the increase in absorbance (and hence the concentration of photoinitiator). The maximum

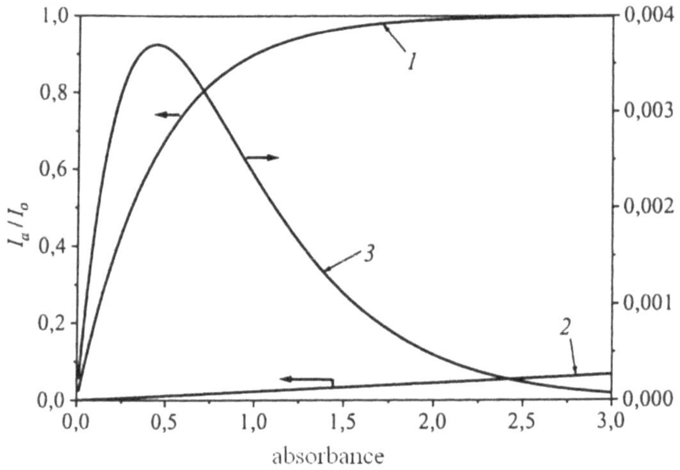

Figure 3.5 Dependence of the ratio I_a/I_0 from the absorbance of the composition (1) throughout the irradiated layer, (2) in the top 1% of the irradiated layer, and (3) in the bottom 1% of the irradiated layer [32].

value of the ratio I_a/I_0 is obtained for $A = 0.43$. It is the optimal value of absorbance, which should characterize the photocurable composites to obtain the maximum absorption of the polymerized sample–base interphase. It is assumed that $A = 0.3$ is an acceptable value [31].

3.2 Conventional Methods of Characterization of Mechanical Properties of Dental Composites

3.2.1 *Microhardness Tests*

Due to the fact that hardness correlates well with other mechanical properties, such as Young's modulus, and the measurements of hardness are relatively easy to conduct, these tests are commonly used to assess the mechanical strength of dental composites. But it should be remembered that hardness is not a material constant that can be clearly defined, but rather a complex property that strongly depends on the method and conditions of measurement.

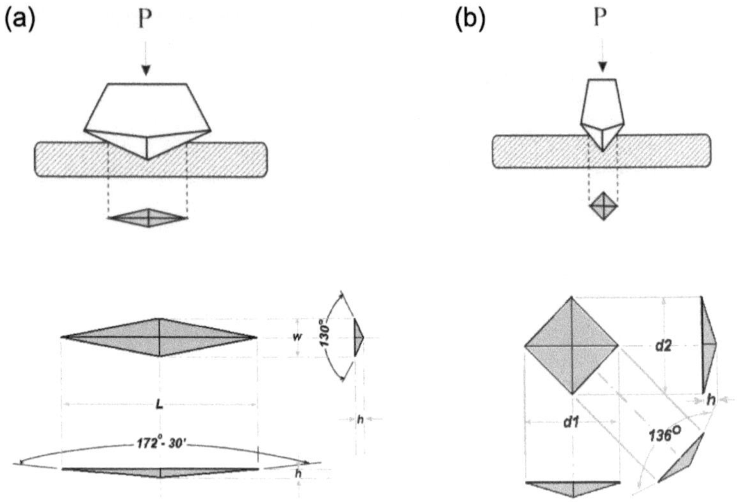

Figure 3.6 Schemes of indenters: (a) Knoop and (b) Vickers. *P*, load applied, usually 0.01–10 N.

In the physical sense, hardness measurements quantify the resistance of a material to plastic deformation. Hardness can be determined on the basis of the simple Eq. 3.4:

$$H = \frac{P}{A_{\text{proj}}} \tag{3.4}$$

where H is the hardness (Pa), P is the load (N), and A_{proj} is the surface area of indent (m^2).

In the context of mechanical properties of light-cured dental composites, hardness measurements are used most frequently to assess the degree of crosslinking. This assessment can be done by measuring the hardness of a material at certain depths [33].

Due to the small size of dental fillings, the hardness measurements of dental composites should be made with small indenters and relatively small loads. For this kind of methods, dentistry uses the term "microhardness" measurements [34].

In the case of microhardness measurement of dental composites, Knoop or Vickers indenters are used most commonly (Fig. 3.6) [18, 33–35].

The difference between the Vickers and Knoop tests is simply the shape of the diamond pyramid indenter. The Vickers test uses a square pyramidal indenter that is prone to cracking of brittle materials. Consequently, the Knoop test using a rhombic-based pyramidal indenter was developed to produce longer but shallower indentations.

During these tests, a small diamond pyramid is pressed into the sample under load. An applied load ranging from 0.01 to 10 N is used. This low amount of load creates a small indent that must be measured under a microscope (by measuring the diagonals of the indent).

Both methods provide high accuracy and allow the use of one scale for the entire range of hardness [36].

3.2.2 *Three-Point Bending Test: The Only Test of Mechanical Properties Included in PN-EN ISO 4049 Standard*

Flexural strength is a mechanical property selected by the International Organization for Standardization (ISO) for screening of resin-based filling materials. The three-point bending test is the only mechanical test included in the standard for dental composites [37].

Flexural strength (σ_g) (MPa) can be calculated on the basis of Eq. 3.5:

$$\sigma g = \frac{3\,Fl}{2\,bh^2} \qquad (3.5)$$

where F is the load at the center of the beam (N), l is the support span (mm), b is the width of the specimen (mm), and h is the thickness of the specimen (mm).

A diagrammatic representation of the flexural test setup is presented in Fig. 3.7.

3.2.3 *Limitation of Conventional Tests*

Microhardness measurements are troublesome and have some limitations, mainly due to the significant influence of human (operator) factors on the obtained numerical values. It seems that accuracy, precision, and reproducibility of results obtained from

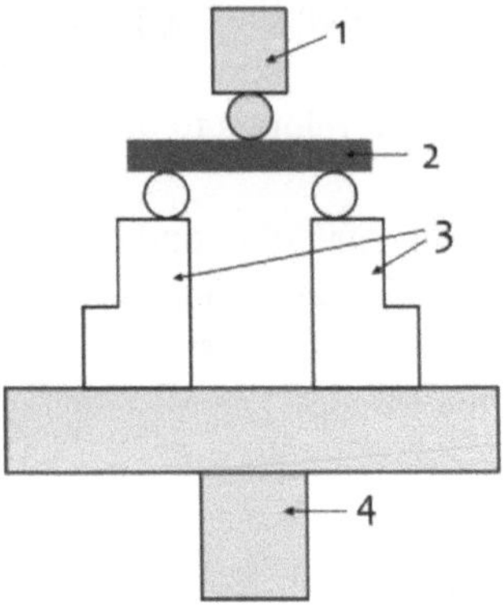

Figure 3.7 Flexural three-point bonding test setup. (1) Loading nose attached to Instron, (2) specimen, (3) lower supports attached to adjustable sliding stage, and (4) lower platform attached to Instron.

microhardness measurements are dependent, in a meaningful way, on the experience of the operator [38].

The static three-point bending test requires the preparation of relatively large samples (25 × 2 × 2 mm) differing greatly in size from the diameter of a dental lamp optical fiber. A disadvantage of this method is that such a long sample requires repeated irradiation, which may result in overlapping of irradiated areas. The reaction between the activator and the photoinitiator, within the overlapping zones of radiation, generates a greater number of radicals, resulting in a higher degree of crosslinking in comparison to neighboring areas and the sample becomes nonhomogeneous. Such heterogeneity may lead to different stress distribution in the material, which is reflected in the varied values of flexural strength. It should also be noted that preparation of such long samples is difficult from a technical point of view and, actually, completely "clinically" unfounded because the real dental fillings are several times smaller [39, 40].

It seems that the three-point bending test, more than for dental composites, is reasonable for determining the flexural strength of fixed partial dentures, which have a similar size like the samples tested. Also, the loads applied in three-point bending tests are similar to those that occur during chewing [18].

3.3 Novel Approach to Dental Composite Characterization

3.3.1 *Properties of the Surface Layer: Nanoindentation Tests*

Hardness may be an indicator of ease of finishing (grinding, polishing) of the surface filling and its resistance to scratches during the use. This is important not only for aesthetic reasons but also for durability, as lower scratch resistance increases the risk of microcrack propagation and can lead to deterioration of mechanical properties of tooth fillings [18]. In this context, it seems reasonable to focus attention primarily on the surface layer, in which the tribological and aging processes are initiated. It seems that it is the quality of this layer that determines the quality of the dental filling.

Nanoindentation is a fast and reliable technique for evaluating local mechanical properties, such as hardness and elastic modulus, of very small volumes of the material. During nanoindentation, a precisely defined three-sided diamond pyramid with an equilateral triangle as the base area is brought into contact with the sample surface. The applied load and the displacement into the specimen are recorded simultaneously. The load–displacement curves obtained during an indentation experiment are essential to calculate the hardness and elastic modulus of the test material.

Nanoindentation is probably the only one technique that allows one to precisely define the mechanical properties of the surface layer [41]. This technique allows obtaining information that would not be provided by volumetric experiments. Probably because of that, this method has gained special importance and popularity in the studies of mechanical properties of hard tissues (including teeth) and biomaterials [42–45].

Nanoindentation is noninvasive and allows one to avoid operator errors, because the whole process is controlled by an automated measuring system. Application of small forces (0.5–5 mN) allows one to explore the surface layer of a material also in the elastic deformation range. The latter differentiates nanoindentation from microhardness measurements, which concern only the plastic deformations of materials.

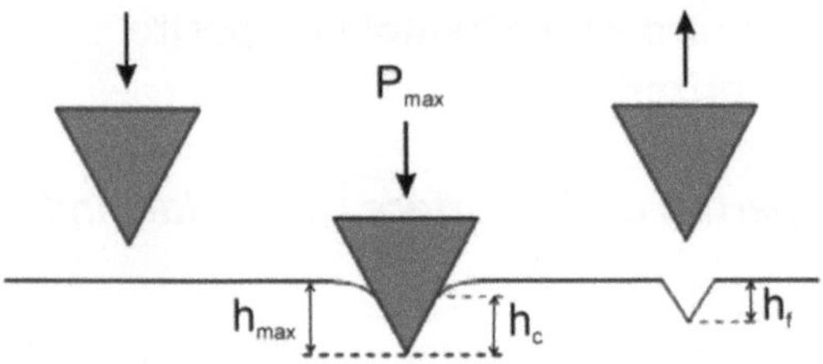

Figure 3.8 Scheme of the indentation process. P_{max}, maximum load applied; h_{max}, penetration depth; h_c, contact depth (height of the contact between the tip and the sample); h_f, final depth.

In a commercial apparatus, the displacement of the indenter is typically monitored by capacitance or inductance (the indenter is mounted directly onto the middle plate of the capacitor), whereas force actuation is provided through electrostatic force generation, magnetic coils, or expansion of a piezoelectric element [42]. The interaction between the tip and the sample during a nanoindentation experiment is illustrated in Fig. 3.8 [42].

Load and displacement are monitored continuously during the experiment, resulting in a load–displacement curve, as shown in Fig. 3.9.

The analytical model proposed by Oliver and Pharr [33], where the unloading curve is subjected to analysis, allows one to determine both the hardness and Young's modulus. Taking into account the

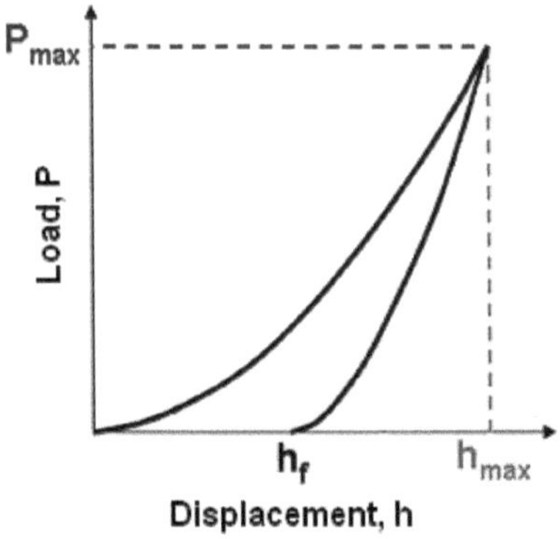

Figure 3.9 Scheme of a typical load–displacement curve.

dimensions of filler particles in dental composites, and dimensions of a tooth filling, it should be noticed that nanoindentation is an ideal technique to evaluate mechanical properties of dental composites [46–49].

According to the literature [46], the reduced modulus[a] of a dimethacrylate-based dental composite is located generally in the range of 15–45 GPa, with nanohardness about 1–2 GPa. For comparison, the nanohardness of human tooth enamel may be twice higher (3.5–4 GPa) with a significantly higher reduced modulus, about 75–80 GPa [43]. When we move toward the dentin (the deeper parts of the tooth), the enamel nanohardness decreases to about 3 GPa and the modulus to about 60 GPa [46], but still these values are several times higher than for composite materials.

Due to the extremely diverse morphology of dental composites and the usage of light curing technique, which may result in a crosslinking gradient, it should be considered whether it is appropriate to describe the mechanical properties by a single, specific numerical value. If we assume that due to the gradient of crosslinking, the top layer of the filling is characterized by increased hardness (in comparison to the interior part of the material) (Fig. 3.10), any generalizations that describe the hardness of a dental material by a specific numerical value become a major misstatement.

3.3.2 *Tribological Experiments*

Wear of materials is a natural process that always occurs when one surface moves in relation to the other, under load. This process is generally considered as negative because of the mass loss and changes of the initial size of materials.

The key factor that determines the amount of wear is the time during which the surfaces are in motion [51]. Healthy enamel of human teeth is durable and it does not generally require any medical intervention for many years. In a sense, the same is expected of dental materials used for the reconstruction of occlusal surfaces

[a]The measured modulus is a function of the contact created by the indenter (usually diamond)-tested material pair. For this reason, the more precise term is "reduced modulus."

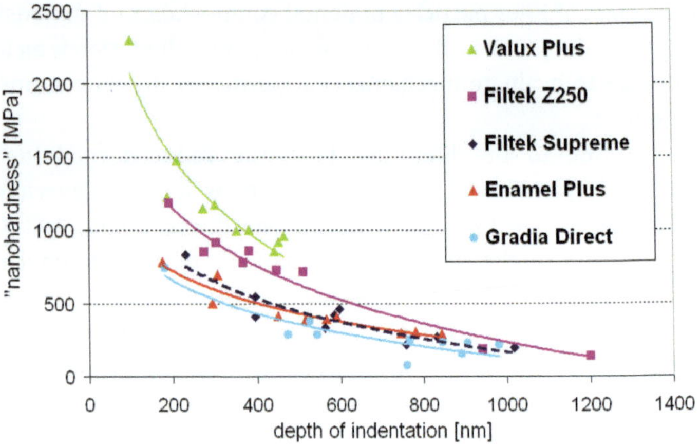

Figure 3.10 Nanohardness profiles of selected dental materials [50].

of teeth. However, to design materials with desired properties, the nature of wear mechanisms of teeth and fillings should be explored first.

We have proposed the analysis of the friction of dental composites using a block-on-ring T-05 tribometer (ITeE, Radom, Poland), with a carbide ring of 35 mm diameter, rotating against a flat block sample of dental composite material. The tribometer should be equipped with a multichannel electronic PC measurement unit enabling data acquisition.

In our study, the friction experiments were run for six hours under a normal load of 100 N. After every hour samples were removed from the instrument and weighed to determine the abrasion of the material. Density of the examined materials was determined by means of double-weighing according to PN-EN ISO 1183-1:2004 (A).

Analysis of the abrasive wear kinetics of commercial dental composites allows observing of some regularity. Some composites exhibit a relatively small material loss in the first hour of the friction test. This may confirm that their surface has a higher resistance to abrasion compared to the bulk of the material (so-called skin effect) [52].

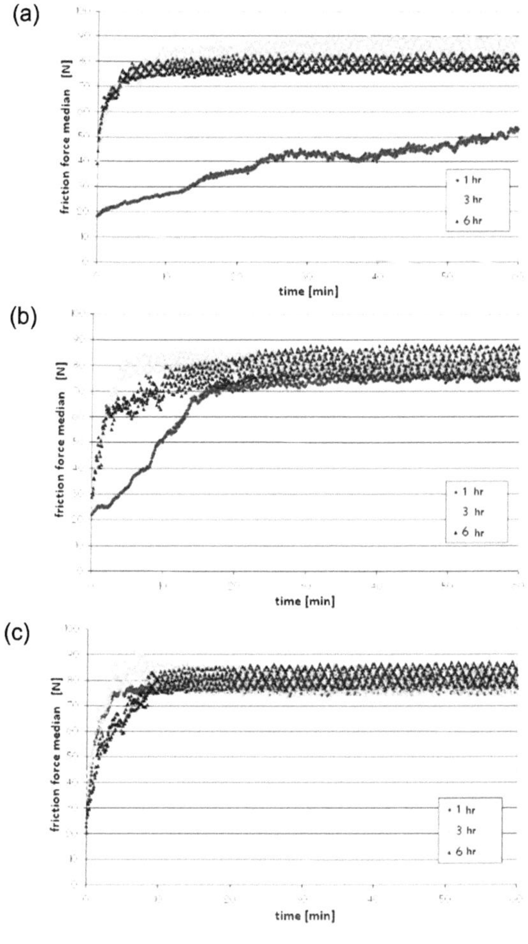

Figure 3.11 Changes to the median of friction force in time for dental composites (a) when friction force equilibrates slowly and (b) when friction equilibrates more rapidly. The value of friction force stabilizes already after 20–25 minutes. (c) Friction presents generally a constant value during the whole experiment.

Similar conclusions can be drawn by analyzing the median value of friction force during the six-hour experiments. Three different friction characteristics of the dental composites are presented in Fig. 3.11.

The slowest equilibration of friction force (Fig. 3.11a) has resulted in a relatively small loss of volume in the first hour of friction. This indicates the presence of the surface layer, which has a slightly different character than the bulk of the material. It results in increased resistance to abrasion of the composite at the initial phase of the tribological experiment. However, the highest overall resistance to abrasion is not connected directly with the wear volume of samples during the first hour of the experiment. However, in the case of dental composites, it should be considered whether the mechanical properties of the surface layer and its resistance to abrasion are not more important than the properties of the bulk of materials.

3.3.3 *Fracture Toughness of Dental Fillings*

In our study, the impact toughness of composites was analyzed using an instrument Resil 5.5 Charpy Hammer (CEAST, Italy), operating with s mass of 0.238 kg and an initial angle of 146°, which gives an impact energy of 0.98 J and a working range of 1.7 kN.

To avoid the necessity to prepare large-sized samples (standard measurements are performed for bars of 17 × 10 × 4 mm dimensions), we have proposed to redesign the sample holder of the instrument. Crushing impact was applied instead of conventional breaking in order to better reflect the real use conditions in the oral cavity. It allowed us to carry out the measurements on small cylinder-shaped samples ($l = 5$ mm, $d = 2$ mm), similar in size to dental fillings (Fig. 3.12). Each time 10 samples were measured, of which 6 with the closest results were taken for further analysis.

The diagrammatic representation of the impact toughness experiment setup is illustrated in Fig. 3.13.

Table 3.2 presents the parameters of impact toughness for the dental composites studied. The magnitude of the median was calculated for maximum dissipated force ($F_{\max}$), the time needed to reach $F_{\max}$ (t), and energy absorption (E) during the impact test. Figure 3.14 presents changes of force and energy versus impact time, registered for the composites studied.

An increase in the filler loading, together with nonuniform size distribution of its particles, makes the energy absorbed by

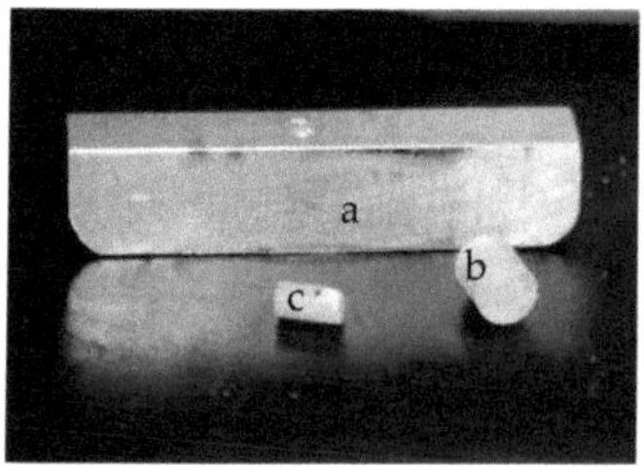

Figure 3.12 (a) Sample holder, (b) polypropylene sample form, and (c) sample for the impact toughness measurements.

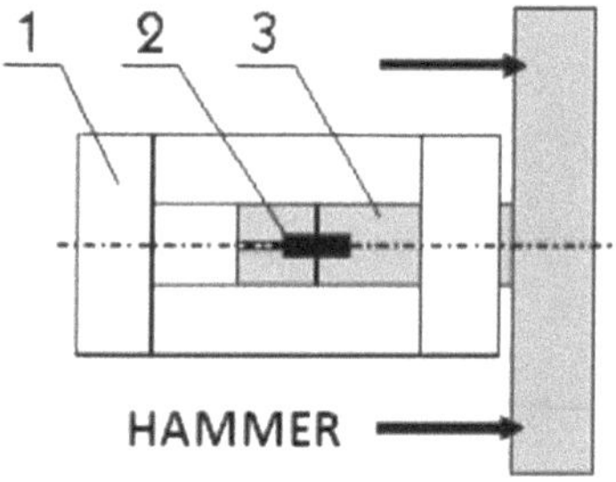

Figure 3.13 Scheme of the sample holder: (1) horizontal frame, (2) sample, and (3) vertical frame [53].

Table 3.2 Impact toughness parameters for the dental composites studied

Material	t(ms)	F_{max} (N)	E (mJ)	Vol.% of filler
QuiXfil (Dentsply DeTrey, Germany)	0.227	413.7	62.3	66
Filtek P60 (3M ESPE, USA)	0.248	389.7	56.7	61
Enamel Plus (Micerium Spa, Italy)	0.257	368.9	55.5	53

the composite material higher, simultaneously shortening the time needed to reach the dissipated force maximum (see data for QuixFil Universal). The difference in energy absorbed by a sample made of Filtek P60 and Enamel Plus materials is insignificant. It can be the result of the same spherical shape and only a slight difference in the size distributions of their hard-phase particles. Varying filler content does not seem to play any role in the energy-absorbing ability of the materials [53].

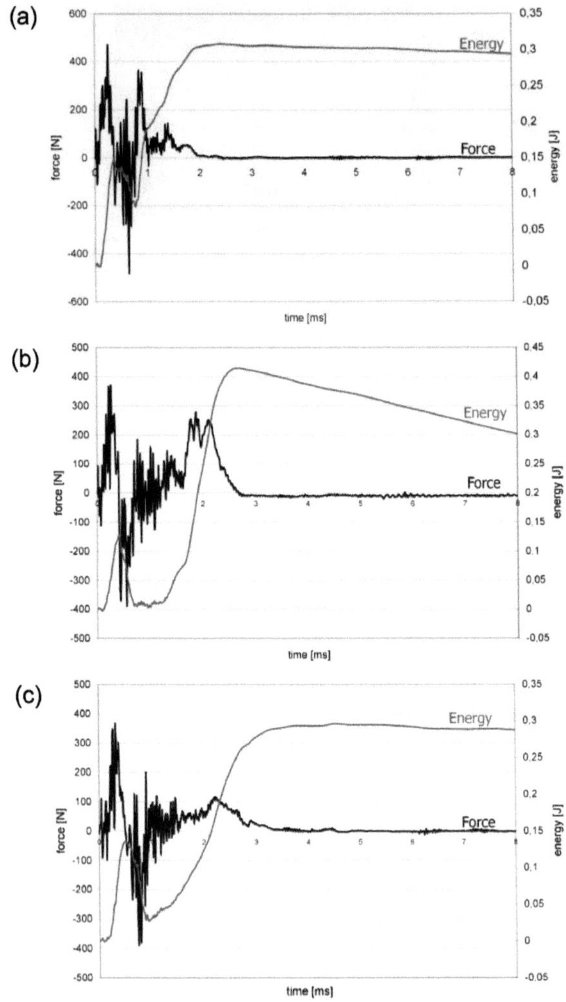

Figure 3.14 Force and energy vs. impact time curves for (a) QuixFil, (b) Filtek P60, and (c) Enamel Plus.

The implementation of 10 repetitions for each material tested showed the existence of considerable discrepancies in numerical values. Such a large diversity of the parameter values pointed out the highly heterogeneous nature of the dental composites (even those from one batch). Even more, it is worth considering the validity of preparation of such large samples (according to the standard [37]) made from more than one syringe of the composite. In fact, each batch of a commercial material may be completely different from others, so the preparation of samples from several batches seems to be completely pointless.

Although the standard for the impact testing of plastics does not include the possibility to perform the tests on samples other than rectangular bars, the solution proposed by us seems to be much more congruent with the nature and specificity of dental fillings. The experiments carried on smaller samples, that is, with similar dimensions to the actual size of a filling, create the possibility of testing a huge number of samples from one batch of composites without wasting large amounts of material. In my opinion, it is justified, not only practically, but also economically.

3.3.4 *Positron Annihilation Lifetime Spectroscopy*

Positron annihilation lifetime spectroscopy (PALS) is a well-established experimental technique for determination of free volume in multicomponent materials and for characterization of their nanoscale porosity.

It is a nondestructive, radioactive technique in which the antiparticles of electron (positrons) from a radioactive source are injected into the studied material. In the case of polymers some of them are trapped in regions of low electron density, which are free-volume holes of 0.2–0.6 nm radius. In this volume the bound state between positron and electron, called positronium (Ps), is formed. From two spin states of Ps only *ortho*-positronium (o-Ps) lives in vacuum for a long time (142 ns), enabling its application in further analysis. o-Ps localized in a free-volume hole annihilates with an electron from molecules of a wall in a significantly shorter time (in the range of several ns). This process is called pick-off annihilation and is the main process of Ps annihilation in polymers.

The lifetime of Ps is related to the size of the free volume, whereas the signal intensity is related to the number concentration of free-volume sites. The lifetime might also depend to some extent on the chemistry of functional groups in the polymer.

Analysis of the positron lifetime spectra of polymers gives usually three (sometimes four) components, attributed to para-Ps annihilation (τ_1), free positron annihilation (τ_2), and o-Ps annihilation (τ_3 and τ_4). The third and fourth components differentiate regions of various material packing density, which are very often present in semicrystalline polymers or composites. The longest lifetimes come from annihilation of o-Ps in the amorphous phase, whereas in crystalline regions they usually do not exceed 1 ns. There is a correlation between the composition of material (filler content) and the intensity of the last two lifetime components. Generally, if the longest positron lifetime is related to the o-Ps annihilation in the polymer matrix, its intensity (associated with the probability of the formation of o-Ps) decreases with increasing filler volume fraction, as a result of the decreasing concentration of free volume in the composite [54–57]. The four components of PALS analysis of polytetrafluoroethylene (PTFE)/silica composites indicates the presence of two distinct distributions of free volumes associated with filler concentration. Their intensities (I_3 and I_4) strongly depend on the concentration of silica. The value of I_3, representing the annihilation in the solid phase, increases, whereas the value of I_4, associated with the polymer matrix, decreases with an increase of filler loading.

However, not only the concentration of the filler may affect the o-Ps lifetime. The shorter annihilation time of o-Ps in plastic/nanosilica composites in comparison to a polymer filled with glass beads was explained by facilitated diffusion of positrons from the filler particles to the polymer matrix [57]. This was confirmed by the increase of intensity of the adequate lifetime component. Intensity of the component originated from the annihilation of o-Ps into the solid phase differs from linear behaviour, for example, due to changes in the morphology/organization of filler particles with concentration.

A conventional *fast-fast* spectrometer with BaF_2 scintillators was used for acquiring positron lifetime spectra. The time resolution (full width at half-maximum [FWHM]) of the system was 250 ps. The positron source, ^{22}Na, of activity 30 μCi, enveloped in 7 μm Kapton foil, was sandwiched between the two samples under investigation. A typical spectrum contained more than 10^6 counts. The obtained spectra were analyzed using the LT routine [58]. The average time of

Table 3.3 Description of the samples for PALS studies

Material	Matrix	Filler	Filler content (wt.%)
Filtek Supreme (3M ESPE, USA)	Bis-GMA, Bis-EMA, UDMA, TEGDMA	nanosilica, SiO_2/ ZrO_2 nanoclusters prepolymerized	78
Gradia Direct (GC, USA)	UDMA	SiO_2, F-Al-Si glass	75
Experimental composite 1	Bis-GMA, TEGDMA	SiO_2, 5 wt.% Si-doped hydroxyapatite	60
Experimental composite 2	Bis-GMA, TEGDMA	SiO_2, nanosilica with nanosilver (~100 ppm)	54

positron annihilation (τ_m) was calculated on the basis of Eq. 3.6:

$$\tau_m = I_1\tau_1 + I_2\tau_2 + I_3\tau_3 + I_4\tau_4 \qquad (3.6)$$

where τ_1, τ_2, τ_3, and τ_4 are components of the positron lifetime and I_1, I_2, I_3, and I_4 are intensities of the components.

Two commercially available resin composite formulations, Filtek Supreme and Gradia Direct, and two homemade composites (Composite 1 and 2) were the objects of examination. The materials were selected for their significantly different morphologies and filler contents (Table 3.3).

The characteristic lifetimes τ_i are calculated as the mass centers of peaks and the intensities I_i as the related areas under the peaks. In the case of the dental composites studied, four peaks appeared in the obtained lifetime distributions. Discrete components of the positron lifetime have been found. The first one, located close to $\tau_1 \approx 130$ ps originates from the annihilation of para-Ps. The second one at $\tau_2 =$ 360–410 ps comes from the annihilation of free positrons. The third one, present at $\tau_3 = 890$–1870 ps, probably comes from the more densely packed regions of the polymer in a composite. The fourth peak, at $\tau_4 = 2110$–3370 ps, corresponds to o-Ps annihilation in the main part of the polymer matrix. The results of PALS analysis are given in Table 3.4.

On the basis of the above analysis (Table 3.4) one can notice that Gradia Direct differs significantly from other composites in the value of I_2. The small deviation from the average value of

Table 3.4 Parameters of PALS analysis for the dental composites studied

Material	τ_2(ns)	I_2(%)	τ_3(ns)	I_3 (%)	τ_4 (ns)	I_4(%)
Filtek	0.410 ± 0.003	60.0 ± 0.9	1.303 ± 0.016	16.0 ± 0.9	2.534 ± 0.071	11.7 ± 0.7
Gradia	0.385 ± 0.011	51.0 ± 1.2	1.008 ± 0.085	18.2 ± 0.9	2.111 ± 0.086	14.7 ± 1.5
Composite 1	0.357 ± 0.010	61.0 ± 1.4	0.893 ± 0.090	12.0 ± 1.1	2.220 ± 0.048	18.7 ± 1.2
Composite 2	0.393 ± 0.004	62.2 ± 3.6	1.867 ± 0.049	22.3 ± 0.9	3.370 ± 0.180	6.1 ± 1.3

τ_2 for the composites studied is also observed for Composite 1. Lower value of the intensity of the second component reflects more difficult conditions for self-annihilation of free positrons and is likely to originate from some morphological constraints present in the highly filled composite. Contrary to other materials studied, Composite 1 represents a morphology that facilitates annihilation of free positrons, for example, due to the decrease of micropore size present in the material.

However, most important and fortunately most distinct are the differences concerning o-Ps annihilation, represented by the third and fourth components. The annihilation taking place in the filler–matrix interphase, being relatively similar between commercial composites, becomes completely different for the homemade materials. Composite 2 is characterized by a much higher concentration of free volumes, additionally of bigger size in comparison not only to Composite 1 but also to other materials studied as well. A similar tendency is followed by free volumes present in the polymer matrix. This time, however, Composite 2 revealed the lowest concentration of free volumes but of the largest size. Composite 1 exhibited the highest concentration of free volumes in a polymer matrix of dimensions similar to the commercial composites.

The smallest value of τ_4, together with the higher value of I_4 obtained for Gradia Direct in comparison to Filtek Supreme, probably results from different polymer resins being applied as composite matrices. The UDMA molecule, which constitutes the matrix of Gradia Direct, is smaller and has a higher concentration of double bonds than bis-GMA, the basic component of the Filtek Supreme matrix [1]. Therefore the former resin exhibits the highest crosslink density and is able to form stronger networks. Some

Table 3.5 Selected properties of dimethacrylate oligomers [10]

Oligomer	M_W	Concentration of double bonds (mol/kg)	Limiting degree of conversion (%)
Bis-EMA	540	3.70	52.2
Bis-GMA	512	3.90	39.0
UDMA	470	4.25	69.6
TEGDMA	286	6.99	75.7

characterization, made for comparison between dimethacrylate resins, is presented in Table 3.5 [10].

The matrix in Composites 1 and 2 is the same—the mixture of bis-GMA and TEGDMA resins. However, a higher value of τ_4 for Composite 2 means probably that the polymer network in this sample has lower crosslink density. Because of the same kind of polymer matrix and the same time of light curing, the difference most likely can be explained by the addition of silver nanoparticles to Composite 2. They worsened the conditions of polymerization by changing the color and transparency of the material, making the polymer matrix less dense.

Reasons for the differences in τ_3 between the composites studied not only are related to filler loading but also depend on the morphology of the materials and interphase interactions. Scanning electron microscopy (SEM) morphology of the composites was analyzed in order to correlate the images with values of the third and fourth components of positron annihilation (o-Ps annihilation) for the materials (Fig. 3.15).

Although the filler loading of Composite 2 is only 54 wt.%, the material exhibits the highest τ_3 and I_3 values. The analysis of SEM images (Fig. 3.15d) does not give any simple explanation. Probably all solid-phase particles in this composite form big, loose agglomerates, infiltrated and separated by tiny layers of polymer. It explains the lowest values of I_4 in this case. Free volumes seldom appear in the polymer matrix, but if they are there, their dimensions are large (the highest value of τ_4).

Images of commercial composites exhibit a significantly different morphology from the experimental ones. The former show bimodal size distribution of filler particles (for Gradia Direct particles

(a) τ_3 =1.30 ns, I_3 =16.0 %, τ_4 =2.53 ns, I_4 =11.7 %

(b) τ_3 =1.01 ns, I_3 =18.2 %, τ_4 =2.11 ns, I_4 =14.7 %

(c) τ_3 =0.89 ns, I_3 =12.0 %, τ_4 =2.22 ns, I_4 =18.7 %

(d) τ_3 =1.87 ns, I_3 =22.3 %, τ_4 =3.37 ns, I_4 =6.1 %

Figure 3.15 SEM cross-section morphology of the composites studied: (a) Filtek, (b) Gradia, (c) Composite 1, and (d) Composite 2 in comparison with the third and fourth components of the positron lifetime.

additionally vary in shape), whereas the latter present solid-phase particles organized in loosely packed big agglomerates. SEM analysis confirms that the degree of particle packing for Composite 1 is significantly higher in comparison to Composite 2 (Fig. 3.15c,d). Also the separation layers between agglomerates are bigger for the latter. This is reflected by the parameters of PALS analysis. Probably the morphology of homemade composites in reality looks like it is presented schematically in Fig. 3.16.

PALS provides information on size and concentration of free volumes in a polymer matrix and in a polymer-filler interphase. On the basis of the analysis it is possible to estimate the crosslink density or compare the quality of polymer-filler interactions.

PALS together with SEM analysis proved to be a useful experimental technique for the determination of dental composites'

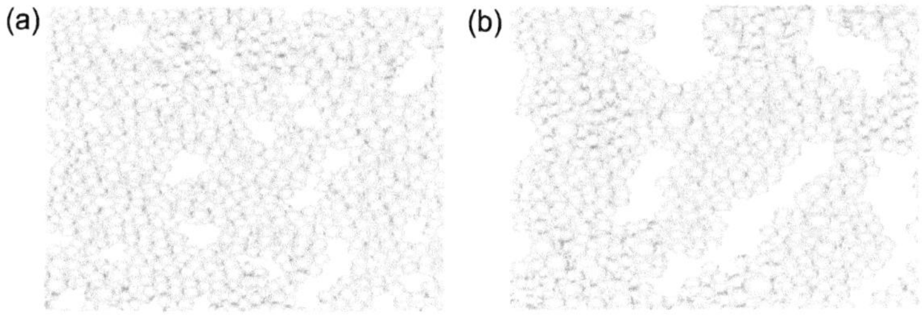

Figure 3.16 Simplified pictures of morphology of homemade composites: (a) Composite 1 and (b) Composite 2.

morphology. On the basis of the analysis of time and intensity of o-Ps annihilation it was possible to differentiate composites, previously selected due to their different mechanical and tribological properties. In my opinion, only the application of both these techniques allows for the precise characterization of the morphology of dental materials with such a high filler content.

3.4 Summary

Over the past years, dimethacrylate-based dental composites for the reconstruction of occlusal surfaces of teeth evolved into a truly satisfactory form. For aesthetic reasons, they leave far behind other types of fillings, like amalgams, but their relatively low mechanical strength, shrinkage during polymerization, and tendency to accumulate dental plaque, resulting in the formation of secondary caries around fillings, still represent significant problems.

Analysis of the subject literature leads to a series of alarming conclusions. Above all, the lack of cooperation between dentistry and materials engineering is visible. Availability of a number of different materials in the dental market, which vary in the quality and content of the inorganic phase, translates into significant differences in their internal structure. The different morphology, in turn, results in different behavior of composites in the oral cavity. The knowledge about the real impact of the size, shape, and quantity of filler particles on the mechanical properties of composites is not systematized. What is more, each batch of a commercial composite is significantly different from another, which makes difficult the formulation of clear conclusions regarding its mechanical properties.

In Part 3 of this chapter, I have proposed a set of tests on samples sized similarly to the dimensions of dental fillings, which seems to be more appropriate than measurements performed according to the ISO 4049 standard "Dentistry. Polymer-Based Filling, Restorative and Luting Materials." PALS supplemented by SEM analysis proved to be a useful experimental technique for the determination of dental composite morphology. On the basis of the analysis of time and intensity of o-Ps annihilation it is possible to differentiate composites in terms of their internal structure.

References

1. Moszner, N., and Salz, U. (2001). New developments of polymeric dental composites, *Prog. Polym. Sci.*, **26**, 535–576.
2. Peutzfeld, A. (1997). Resin composites in dentistry: the monomer systems, *Eur. J. Oral Sci.*, **105**, 97–116.
3. Braga, R.R., Ballester, R.Y., and Ferracane, J.L. (2005). Factors involved in the development of polymerization shrinkage stress in resin-composites: a systematic review, *Dent. Mater.*, **21**, 962–970.
4. Morgan, R., Kalachandra, S., Shobha, H.K., Gunduz, N., and Stejskal, E.O. (2000). Analysis of a dimethacrylate copolymer (Bis-GMA and TEGDMA) network by DSC and ^{13}C solution and solid-state NMR spectroscopy, *Biomaterials*, **21**, 1897–1903.
5. Albers, H.F. (2002). *Tooth-Colored Restoratives. Principles and Techniques* (BC Deker, Ontario, Canada).
6. Bowen, R.L. (1962). Dental filling material comprising vinyl silane treated fused silica and a binder consisting of reaction product of bis phenol and glycidyl acrylate (US Patent No. 3066112).
7. Foster, J., and Walker, R. J. (1974). Dental filling materials (US Patent No. 3825518).
8. Barszczewska-Rybarek, I. (2008). Badanie wpływu struktury uretano-dimetakrylanów na morfologię i właściwości otrzymywanych z nich polimerów, *Polimery*, **53**, 190–194.

9. Asmussen, E., and Peutzfeldt, A. (1998). Influence of UEDMA, BisGMA and TEGDMA on selected mechanical properties of experimental resin composites, *Dent. Mater.*, **14**, 51–56.
10. Sideridou, I., Tserki, V., and Papanastasiou, G. (2002). Effect of chemical structure on degree of conversion in light-cured dimethacrylate-based dental resins, *Biomaterials*, **3**, 1819–1829.
11. Weinmann, W., Thalacker, C., and Guggenberger, R. (2005). Siloranes in dental composites, *Dent. Mater.*, **21**, 68–74.
12. Ilie, N., Jelen, E., Clementino-Luedemann, T., and Hickel, R. (2007). Low-shrinkage composite for dental application, *Dent. Mater. J.*, **26**, 149–155.
13. Eick, J.D., Kotha, S.P., Chapelow, C.C., Kilway, K.V., Giese, G.J., Glaros, A.G., and Pinzino, C.S. (2007). Properties of silorane-based dental resins and composites containing a stress reducing monomer, *Dent. Mater.*, **23**, 1011–1017.
14. Lien, W., and Vandewalle, K.S. (2010). Physical properties of a new silorane-based restorative system, *Dent. Mater.*, **26**, 337–344.
15. Klapdohr, S., and Moszner, N. (2005). New inorganic components for dental filling composites, *Monatsh. Chem.*, **136**, 21–45.
16. Rüttermann, S., Krüger, S., Raab, W.H.-M., and Janda, R. (2007). Polymerization shrinkage and hygroscopic expansion of contemporary posterior resin-based filling materials: a comparative study, *J. Dent.*, **35**, 806–813.
17. Stansbury, J.W., Dickens, B., and Liu, D.-W. (1995). Preparation and characterization of cyclopolymerizable resin formulations, *J. Dent. Res.*, **74**, 1110–1115.
18. Craig, R.G. (2006). *Materiały Stomatologiczne*, Powers, J.M., and Sakaguchi, R.L., eds. (Elsevier Urban and Partner, Wrocław).
19. Chan, D.C., Titus, H.W., Chung, K.-H., Dixon, H., Wellinghoff, S.T., and Rawls, H.R. (1999). Radiopacity of tantalum oxide nanoparticle filled resins, *Dent. Mater.*, **15**, 219–222.
20. Söderholm, K-J.M., Lambrechts, P., Sarrett, D., Abe, Y., Yang, M.C.K., Labella, R., Yildiz, E., and Willems, G. (2001). Clinical wear performance of eight experimental dental composites over three years determined by two measuring method, *Eur. J. Oral Sci.*, **109**, 273–281.
21. Xu, H.H.K., Martin, T.A., Antonucci, J.M., and Eichmiller, E.C. (1999). Ceramic whisker reinforcement of dental resin composites, *J. Dent. Res.*, **78**, 706–712.

22. Xu, H.H.K., Quinn, J.B., Smith, D.T., Antonucci, J.M., Schumacher, G.E., and Eichmiller, F.C. (2002). Dental resin composites containing silica-fused whiskers: effects of whisker-to-silica ratio on fracture toughness and indentation properties, *Biomaterials*, **23**, 735–742.
23. Xu, H.H.K., Quinn, J.B., Smith, D.T., Giuseppetti, A.A., and Eichmiller, F.C. (2003). Effects of different whiskers on the reinforcement of dental resin composites, *Dent. Mater.*, **19**, 359–367.
24. Tian, M., Gao, Y., Liu, Y., Liao, Y., Hedin, N.E., and Fong, H. (2008). Fabrication and evaluation of Bis-GMA/TEGDMA dental resins/composites containing nano fibrillar silicate, *Dent. Mater.*, **24**, 235–243.
25. Karbhari, V.M., and Strassler, H. (2007). Effect of fiber architecture on flexural characteristics and fracture of fiber-reinforced dental composites, *Dent. Mater.*, **23**, 960–968.
26. Garoushi, S., Vallittu, P.K., Watts, D.C., and Lassila, L.V.J. (2008). Effect of nanofiller fractions and temperature on polymerization shrinkage on glass fiber reinforced filling materials, *Dent. Mater.*, **24**, 606–610.
27. Sajewicz, D. (2006). On evaluation of wear resistance of tooth enamel and dental materials, *Wear*, **260**, 1256.
28. Bieliński, D.M., Głąb, P., and Ślusarski, L. (2001). FT-IR ATR spectra and nanohardness measurements as used to follow the progress in photocuring of polyester resins, *Polimery*, **7/8**, 494.
29. Bieliński, D.M. (2004). On importance of the surface layer for exploitation of polymer materials, *Kautsch. Gummi Kunstst.*, **57**, 13.
30. Rånby, B., and Rabek, J.F. (1975). *Photodegradation, Photo-oxidation and Photostabilization of Polymers* (John Wiley and Sons, London).
31. Rabek, J.F. (1987). *Mechanisms of Photophysical Processes and Photochemical Reactions in Polymers* (Wiley, Chichester).
32. Pączkowski, J. (2003). *Fotochemia polimerów. Teoria i Zastosowanie* (Wydawnictwo Uniwersytetu Mikołaja Kopernika, Toruń).
33. Rode, K.M., de Freitas, P.M., Lloret, P.R., Powell, L.G., and Turbino M.L. (2009). Micro-hardness evaluation of a micro-hybrid composite resin light cured with halogen light, light-emitting diode and argon ion laser, *Lasers Med. Sci.*, **24**, 87–92.
34. Ryge, G., Foley, D.E., and Fairhurst, C.W. (1961). Micro-indentation hardness, *J. Dent. Res.*, **40**, 1116–1126.
35. Poskus, L.T., Placido, E., and Capel Cardoso, P.E. (2004). Influence of placement techniques on Vickers and Knoop hardness of class II composite resin restorations, *Dent. Mater.*, **20**, 726–732.

36. Dobrzański, L.A. (2002). *Podstawy Nauki o Materiałach i Metaloznawstwo* (WNT Warszawa).
37. PN-EN ISO 4049 Standard (2003). *Stomatologia. Polimerowe Materiały do Wypełnień, Odbudowy i Cementowania.*
38. McGhee, D. (2005). Common problems in microhardness testing, *Heat Treat. Prog.*, **5**, 92–96.
39. Chung, S.M., Yap, A.U.J., Chandra, S.P., and Lim, C.T. (2004). flexural strength of dental composite restoratives: comparison of biaxial and three-point bending test, *J. Biomed. Mater. Res.*, **71B**, 278–283.
40. Yap, A.U.J., and Teoh, S.H. (2003). Comparison of flexural properties of composite restoratives using the ISO and mini-flexural tests, *J. Oral Rehab.*, **30**, 171–177.
41. Oliver, W.C., and Pharr, G.M. (1992). An improved technique for determining hardness and elastic modulus using load and displacement indentation experiments, *J. Mater. Res.*, **7**, 1564–1583.
42. Ebenstein, D.M., and Pruitt, L.A. (2006). Nanoindentation of biological materials, *Nano Today*, **1**, 26–33.
43. Habelitz, S., Marshall, Jr, G.W., Balooch, M., and Marshall, S.J. (2002). Nanoindentation and storage of teeth, *J. Biomech.*, **35**, 995–998.
44. Zysset, P.K., Guo, X.E., Hoffler, C.E., Moore, K.E., and Goldstein, S.A. (1999). Elastic modulus and hardness of cortical and trabecular bone lamellae measured by nanoindentation in the human femur, *J. Biomech.*, **32**, 1005–1012.
45. He, L.-H., and Swain, M.V. (2009). Enamel: a functionally graded natural coating, *J. Dent.*, **37**, 596–603.
46. Drummond, J.L. (2006). Nanoindentation of dental composites, *J. Biomed. Mater. Res.*, **78B**, 27–34.
47. Chung, S.M., Yap, A.U.J., Tsai, K.T., and Yap, F.L. (2004). Elastic modulus of resin-based dental restorative materials: a microindentation approach, *J. Biomed. Mater. Res.*, **72B**, 246–253.
48. He, L.-H., and Swain, M.V. (2007). Nanoindentation derived stress-strain properties of dental materials, *Dent. Mater.*, **23**, 814–821.
49. Masouras, K., Akhtar, R., Watts, D.C., and Silikas, N. (2008). Effect of filler size and shape on local nanoindentation modulus of resin-composites, *J. Mater. Sci. Mater. Med.*, **19**, 3561–3566.
50. Kleczewska, J., Bieliński, D.M., Sokołowski, J., and Klimek, L. (2007). Analiza warstwy wierzchniej polimerowych materiałów dentystycznych, *Inż. Mater.*, **160**, 930–934.

51. Mair, L.H., Stolarski, T.A., Vowles, R.W., and Lloyd, C.H. (1996). Wear: mechanisms, manifestation and measurement. Report of workshop, *J. Dent.*, **24**, 141–148.
52. Bieliński, D.M., Ślusarski, L., Affrossman, S., O'Neil, S.A., and Pethrick, R.A. (1997). Influence of iodination on tribological properties of acrylonitrile–butadiene rubber, *J. Appl. Polym. Sci.*, **64**, 1927–1936.
53. Kleczewska, J., Bieliński, D.M., and Morawiec, J. (2008). Impact toughness of polimer dental composites based on dimethacrylates resin, *Mod. Polym. Mater. Environ. Appl.*, **3**, 49–54.
54. Winberg, P., Eldrup, M., and Maurer, F.H.J. (2004). Nanoscopic properties of silica filled polydimethylsiloxane by means of positron annihilation lifetime spectroscopy, *Polymer*, **45**, 8253–8263.
55. Wang, Y.-Q., Wu, Y.-P., Zhang, H.-F., Zhang, L.-Q., Wang, B., and Wang, Z.-F. (2004). Free volume of montmorillonite/styrene-butadiene rubber nanocomposites estimated by positron annihilation lifetime spectroscopy, *Macromol. Rapid Commun.*, **25**, 1973–1978.
56. Zhang, M., Fang, P.F., Zhang, S.P., Wang, B., and Wang, S.J. (2003). Study of structural characteristics of HDPE/$CaCO_3$ nanocomposites by positrons, *Radiat. Phys. Chem.*, **68**, 565–567.
57. Lizama, B., Castañeda, D., López-Castañares, R., Castaño, V., and Vazquez, F. (2000). Characterization of composite latex particles by positron annihilation spectroscopy, *Macromol. Symp.*, **151**, 419–425.
58. Kansy, J. (1996). Microcomputer program for analysis of positron annihilation lifetime spectra, *Nucl. Instrum. Methods Phys. Res., Sect. A*, **374**, 235–244.

Chapter 4

Properties of a Lead-Free Solder

Ze Ma,[a] Florent Chalon,[a] René Leroy,[a] Denis Bertheau,[b] and Naryanaswami Ranganathan[a]

[a]*Laboratoire de Mécanique et de Rhéologie (LMR) EA2640, Université François Rabelais de Tours (UFRT), Polytech 'Tours, 7 avenue Marcel Dassault, Tours, 37200, France*

[b]*Institut P', ISAE-ENSMA, UPR CNRS 3346, Material Engineering Department, 1 avenue Clément Ader, Téléport 2, BP 40109, F86961 Futuroscope Chasseneuil Cedex, France*

ranganathan@univ-tours.fr

4.1 Introduction

In microelectronics development, flexible electronics have been around for years. For today's design engineer, electronic component flexibility would be a chip that can bend and stretch like human skin. It can be implanted as a health monitor, used in mobile communication systems, which could be integrated into clothing, etc. All these applications demand adaptable packaging. Connection joints take an important role in packaging. The Sn–Ag–Cu lead-free-type alloy is the most promising replacement for lead-based ones and the study of its behavior under different loading situations is therefore critical [1]. Its conventional mechanical properties and also fatigue behaviors are subjects frequently studied [2–8]. In this

Materials Characterization: Modern Methods and Applications
Edited by Narayanaswami Ranganathan

ISBN 978-981-4613-06-4 (Hardcover), 978-981-4613-07-1 (eBook)
www.panstanford.com

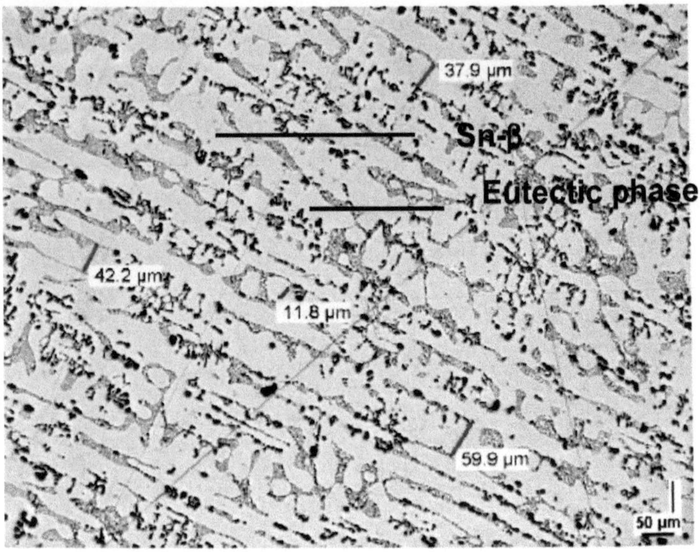

Figure 4.1 Microstructure of a bulk specimen.

chapter, the mechanical properties and fatigue behavior of a Sn–Ag–Cu(Ni) alloy are studied in both global and local scales.

The material's microstructure of the bulk specimen (Fig. 4.1) is observed by an optical microscope after mechanical polishing up to 1 μm. The material is composed of Sn-β and the eutectic phase, as shown in Fig. 4.1.

4.1.1 *Global Properties by Conventional Tests*

The specimens for conventional tests from the as-received alloy are molded under the temperature of fabrication. Its geometry is shown in Fig. 4.2. The experiments are carried out by the DMA Metravib VA 2000 apparatus. Tensile tests were carried out under different temperatures and the results are summarized in Figs. 4.3a and 4.3b. According to the tensile tests, temperature has a strong effect on the mechanical properties of the material. Yield stress, ultimate tensile strength, and modulus decrease linearly with temperature but follow different slopes. As the temperature increases from 20°C

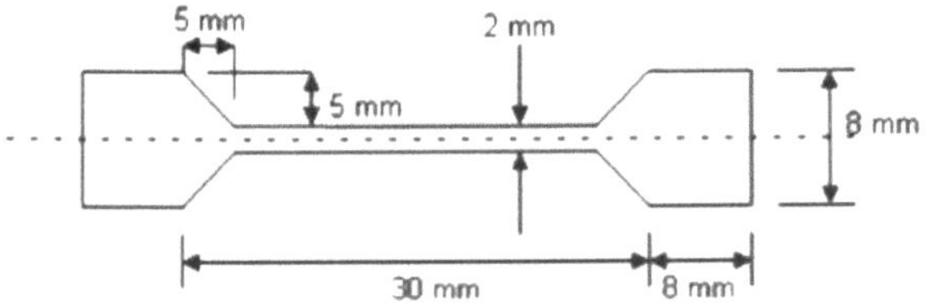

Figure 4.2 Bulk specimen geometry.

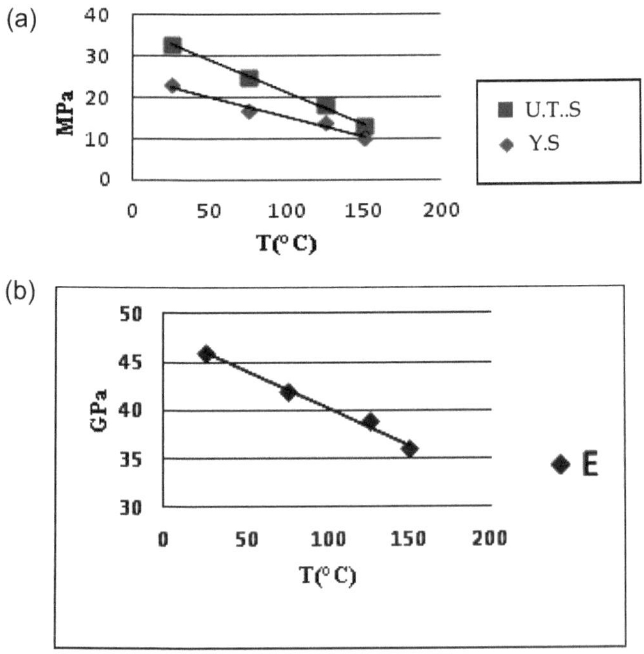

Figure 4.3 (a) Evolutions of yield stress and ultimate tensile strength and (b) evolution of Young's modulus with temperature.

to 150°C, the yield stress and the ultimate tensile stress decrease by about 40%, while the decrease in modulus is about 70%.

As the material has also a low melting point, of 237°C, the material creeps even at room temperature. So understanding creep behavior is essential. The temperature range for creep tests is from 50°C to 150°C. During the test, the material enters the stable secondary phase in a short time. The secondary creep represents

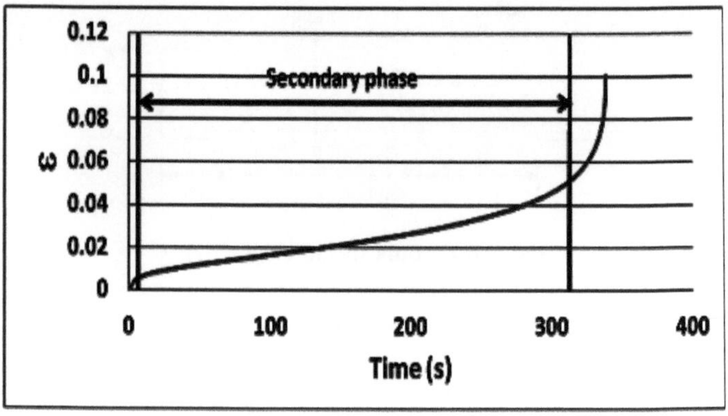

Figure 4.4 Creep experiment 100°C at a stress of 11 MPa.

Table 4.1 Norton law parameters

	A	n	Q_a/k(K)
Norton	8.10^{-5}	9.5	7983

most of the life as compared to the first and third creep phases (Fig. 4.4). Norton's law (Eq. 4.1) is chosen to describe the secondary creep behavior for this material and its parameters are identified and shown in Table 4.1.

$$\varepsilon'_{\text{II}} = A * \sigma^n * \exp\left(-\frac{Q_a}{T * k}\right) \tag{4.1}$$

A is a material constant, n is the stress exponent rate, Q_a is the activation energy, and k is the Boltzmann constant (1.38×10^{-23} J K^{-1}).

4.1.2 *Local Properties*

Nanoindentation is a technique that permits characterization locally at a micron/submicron scale, for example, the properties of a grain or that of the grain boundary. With nanoindentation, both elastic and creep properties can be investigated.

Table 4.2 Elastic modulus and hardness for the two groups of tests

H (GPa), g1	E (GPa), g1	H (GPa), g2	E (GPa), g1
0.2 ± 0.01	47	0.225 ± 0.015	49

The reduced modulus can be directly obtained by nanoidentation tests. The values obtained from two groups of tests are given in Table 4.2.

From conventional tests, the E value at room temperature is about 47 GPa. Thus we find that E values measured from nanoindentation are quite comparable to those obtained from conventional tests.

The secondary creep properties of metals' characterization by nanoindentation tests has been studied by Goodall and Clyne [9]. Unfortunately, creep behavior study using nanoindentation cannot be applied on all kinds of metals, but it remains a reliable method for low-melting-point metals [9–14]. For the material of the present study, the melting point is low that the creep appears at room temperature. The methodology used for creep tests with a Berkovich indenter are described in Chapters 1 and 2 of this book. Basically the creep rate is measured during the hold period and the stresses are measured using the contact area under the applied load

The nano tests were carried out by the NanoTest system from Micro Materials Limited, in a temperature- and humidity-controlled (strictly) room. The temperature was about 25°C.

Two groups of tests have been done—one with a 0.3 mN/s loading rate (g1) and the other with 0.5 mN/s (g2). For the bulk specimen, the stress-sensibility-to-strain rate is obtained by plotting the stress versus strain rate, and a value of is 0.13 is obtained—see Fig. 4.5—and this value is comparable to that obtained from a conventional creep test, which is 0.1.

Creep tests were then carried out in an in situ solder ball, under the same test conditions as for the bulk material.

The tests show that the strain rate sensitivity is quite different and a value of 0.079 is obtained, significantly lower than that obtained for the bulk material (Fig. 4.6).

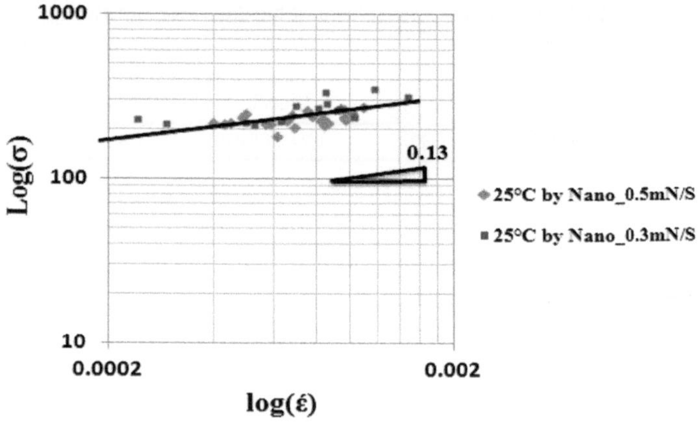

Figure 4.5 Strain rate sensibility measured by nanoindentation.

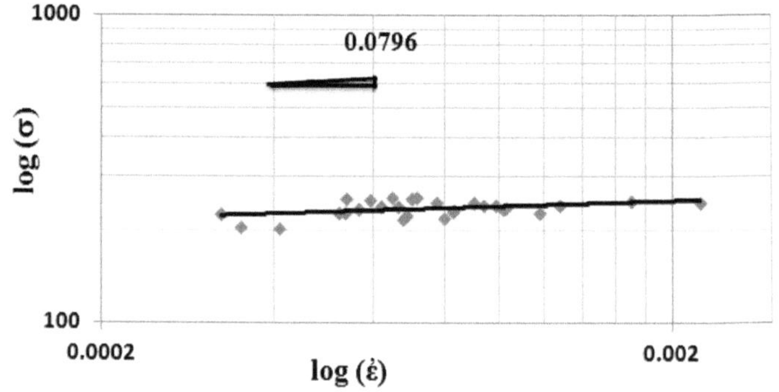

Figure 4.6 Strain rate sensitivity for an in situ solder, measured by nanoindentation.

The modulus values from in situ solder tests are given in Table 4.3. Comparing Tables 4.2 and 4.3 we see that the in situ solder has a hardness about 25% higher than the bulk alloy.

The difference in the mechanical properties of the in situ solder and the bulk material is attributed to differences in microstructure and the presence of precipitates in the in situ solder [15].

Figure 4.7 shows very low load indents in the in situ solder ball. In Fig. 4.8, it can be seen that far from the base, the material has a modulus of 46 GPa, quite similar to that of the bulk alloy. But very

Table 4.3 Elastic modulus of an in situ solder by nanoindentation

E (GPa)	H (GPa)
60.5 ± 4	$0{,}239 \pm 0.04$

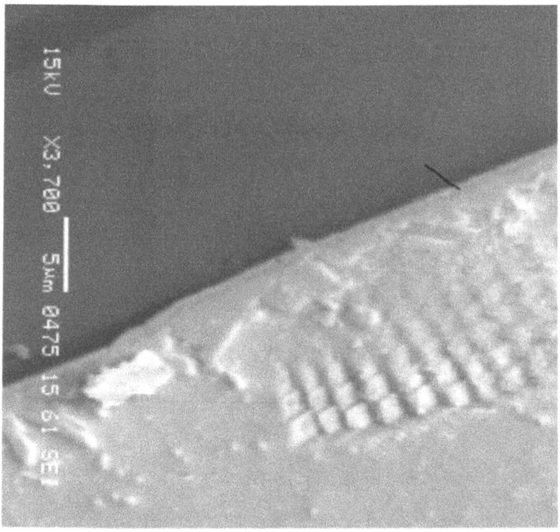

Figure 4.7 Microstructure and indents in the solder ball in situ.

close to the base, where one can see the presence of precipitates, the modulus increases sharply.

In a companion study, the heterogeneous structure and properties of the in situ solder are discussed in detail [15].

4.2 Fatigue Behavior

In flexible chips' applications, the solder joints must tolerate a predeformation and it may influence the durability and reliability of a component containing a lead-free solder. In recent research on lead-free solders, low-cycle fatigue studies are popular. For low-cycle fatigue, the plastic strain range or plastic work is considered the most representative factor that influences the fatigue life. Scientists

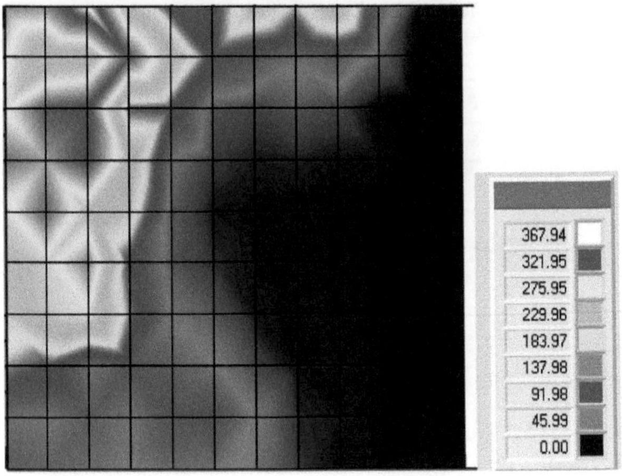

Figure 4.8 Hardness maps in the solder ball in situ.

also work on different parameters that influence the fatigue life, for example, temperature, frequency, and even corrosion effects. Frequency influences the crack initiation and propagation lives. Fatigue ductility increases at low frequencies [8]. Temperature enhances the creep behavior during the fatigue phenomena.

Intergranular cracks are observed on the fracture surface [4, 16]. John et al. [2] have studied the temperature and frequency effects on a Sn–Cu solder and added the frequency effect in a modified Coffin–Manson model. There are also studies on additional element effects such as Bi and Ni that can improve the fatigue life [3, 4]. Though lots of works concerned strain-controlled fatigue experiments, the strain ratio R (ratio between minimum and maximum strains in a cycle), a parameter that can describe predeformation, is not well documented. Lin and Huang [6] have studied Sn–Ag–Cu solder fatigue behavior for load ratios of –1, 0, and 0.5. They found the increasing strain ratio causes a significant reduction of fatigue life due to the mean strain effect as against the mean stress effect. They also modified the Coffin–Manson model to include the effect of load ratio R. These conclusions are important for understanding the predeformation effect. Local fatigue behavior can help to understand the fatigue behavior at a microscopic level. In

Table 4.4 Conventional fatigue tests (total strain values, ε_T, in %)

R_d	Frequency	$\varepsilon_T = 0.27$	$\varepsilon_T = 0.4$	$\varepsilon_T = 0.44$	$\varepsilon_T = 1$	$\varepsilon_T = 2$	$\varepsilon_T = 4$
−1	0.01				X	x	x
0.1	1	x	x	x			
0.5	1	x	x	x			
0.7	1	x	x				
0.9	1	x	x	x			

this study, conventional fatigue tests are complemented with a novel local fatigue investigation method.

4.2.1 *Global Fatigue Behavior: Predeformation Effect on Fatigue*

Displacement-controlled fatigue tests were performed. A range of R_d values was considered for this work. R_d is the ratio of minimum to maximum displacement. These values and the tested strain ranges are given in Table 4.4.

All the tests were carried out at room temperature on smooth, flat specimens using an electrical motor servo-controlled test system operating under displacement control.

4.2.2 *Analysis of Hysteresis Behavior*

A history plot of the stress–strain hysteresis loops for $R_d = -1$ at the 10th, 50th, 100th, 200th, 300th, 400th, and 500th cycles tested at 4% total strain range is given in Fig. 4.9a. The stress amplitude varies with the loading cycles, as shown in Fig. 4.9b. As the stress amplitude decreases during the fatigue test, the material exhibits cyclic softening [17–22]. Different fatigue stages can be identified by stress amplitude evolution. The first stage of fatigue behavior results from cyclic softening, crack initiation, and the coalescence process [17, 18]. The second stage shows a steady reduction in stress amplitude with cycles. The stress amplitude decreases almost linearly with cycles during this second stage. This could be associated with a stable fatigue crack propagation process in the second stage [3, 23]. The final stage shows rapid reduction in

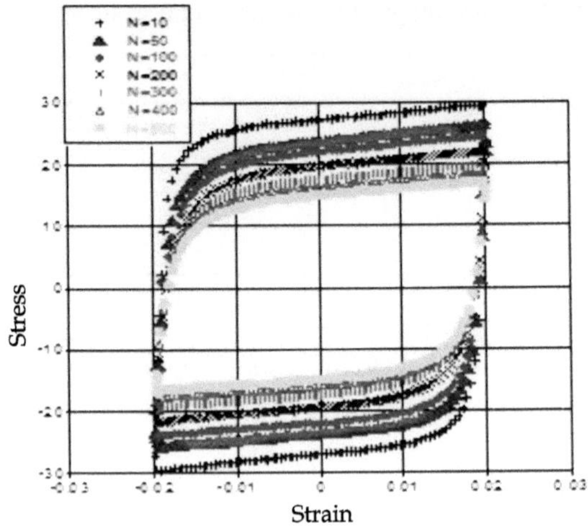

(a) Hysteresis curves of total strain 4% R_d = –1

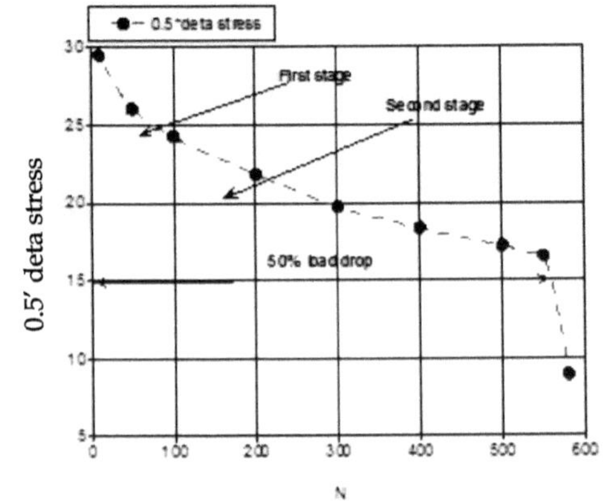

(b) Stress amplitude evolution of total strain R_d = –1

Figure 4.9 Stable hysteresis curves plotting and stress amplitude evolution.

stress amplitude, leading to final ductile rupture of the specimen. When the maximum tensile load level reaches 50% of the initial level, the corresponding number of cycles is reported as the fatigue life.

For different R_d values, the stress amplitude evolutions are compared in Fig. 4.10. According to these results several observations can be drawn:

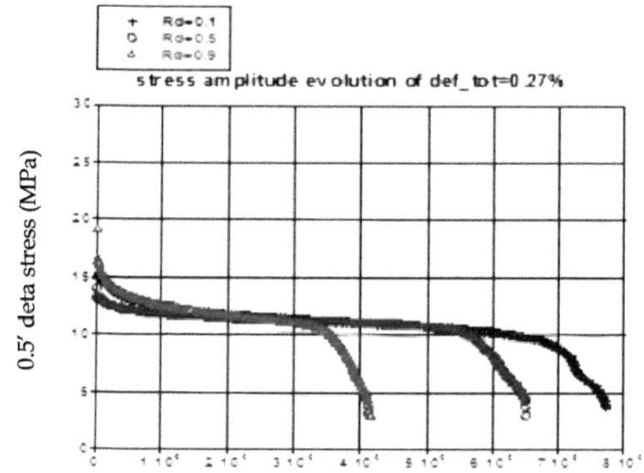

(a) Stress amplitude evolution total strain 0.27%

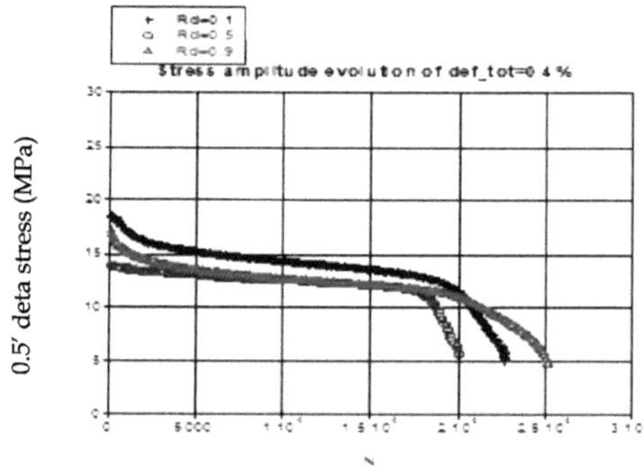

(b) Stress amplitude evolution total strain 0.4%

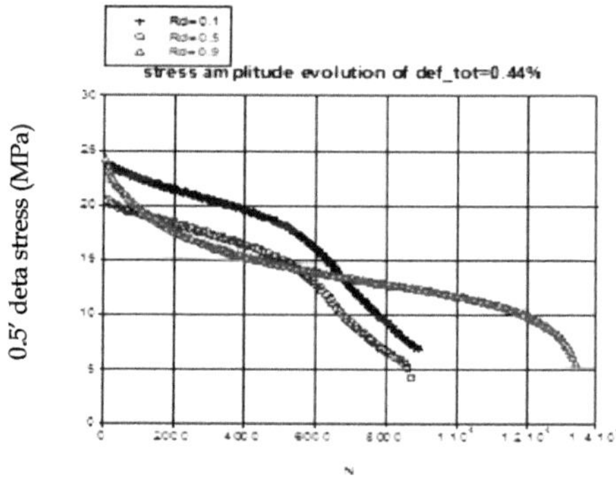

(c) Stress amplitude evolution total strain 0.44%

Figure 4.10 Stress amplitude evolution in different conditions.

- The softening behavior is not affected by R_d.
- The R_d value influences the stable fatigue crack propagation stage. It decreases with an increase in R_d.
- In the low R_d case, the plastic damage dominates and the crack propagation stage depends strongly on strain amplitude. The stronger the solicitation is, the faster the crack propagates, and the shorter the crack propagation stage is. This is the reason why the fatigue life decreases dramatically when the total strain increases and this is similar to the conclusion of Lin [6].
- Compared to a low R_d case, a high R_d value has a weak influence on fatigue life but a strong effect on the first fatigue stage. It is suggested here that under such conditions the dominant damage is creep.

The mean stress evolution during strain cycling with different R_d and total strain values is shown in Fig. 4.11. For different conditions, the mean stress decreases at first cycles and then remains constant. For the $R_d = 0.1$ condition, Fig. 4.11a, the mean stress first increases and then decreases with total strain. But for a high R_d condition, Fig. 4.11b, the mean stress decreases continuously with increasing total strain; this stress relaxation could be correlated to creep.

The stable hysteresis curves of different R_d conditions are plotted in Fig. 4.12. When R_d is smaller than 0.6, the material is hardening. When R_d gets over 0.6, the material is softening, even though the strain amplitude increases. This is caused by the creep effect. The creep enhances with a high R_d value. When R_d crosses 0.6, the creep is strong and a larger displacement can be reached with a smaller load. A displacement ratio of 0.6 can be considered to be a critical value, lower than that which the plastic damage dominates, and for higher R_d values the creep damage is dominant.

The cyclic behavior of metals can be described using the Masing model [23]. Here is a short introduction to Masing behavior. A material that follows the Masing model would have identical loading and unloading branches of the hysteresis loops, and their elastic regions remain unaltered with a change in strain amplitude. Experimentally, Masing behavior in a material is verified by bringing

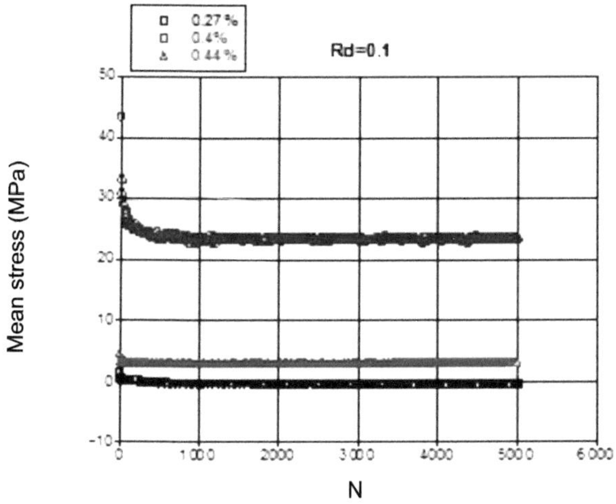

(a) Mean stress evolution of different total strain values $R_d = 0.1$

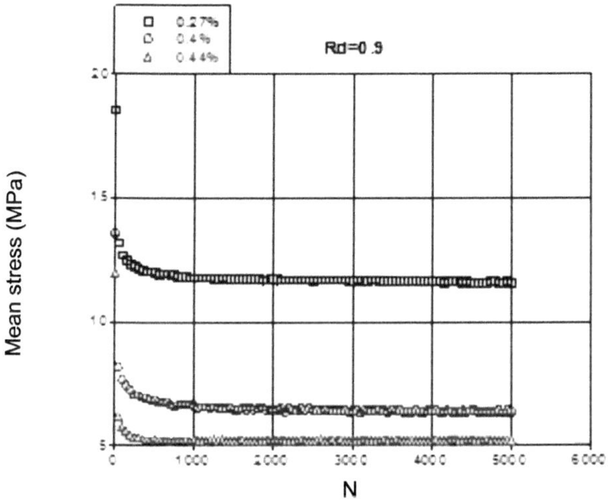

(b) Means stress evolution of different total strain values $R_d = 0.9$

Figure 4.11 Mean stress evolution in different conditions.

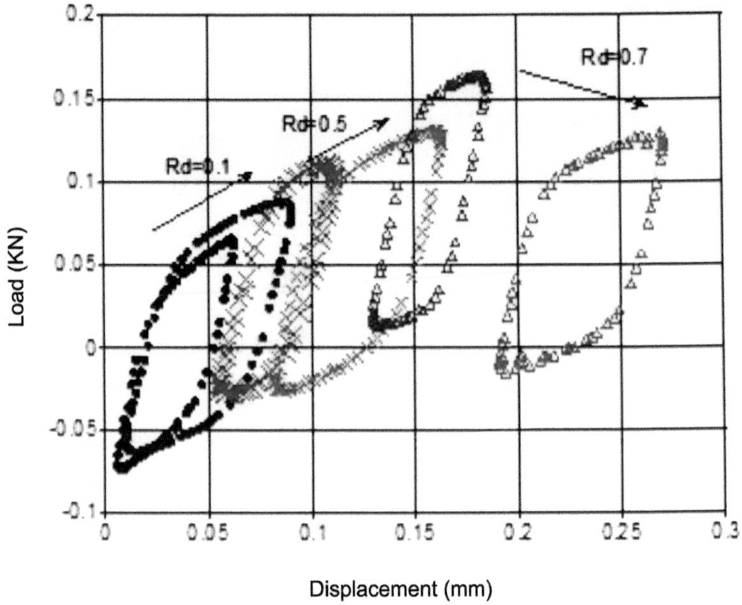

Figure 4.12 Stable hysteresis curves with different R_d conditions.

the compressive tips (sometimes the tensile tips) of all the stable hysteresis loops belonging to various strain amplitudes to a common origin. If all the loading branches (or unloading branches in the case of translating tensile tips) overlap and form a common envelope curve, termed as "master curve," then the material is said to follow the Masing model [23].

The stabilized hysteresis loops of R_d −1, 0.1, 0.9 tests were shifted in a way that the lower tips of the loops moved to the origin, as shown in Fig. 4.13.

We notice that the Masing stress response depends on the total strain range.

These results are now further analyzed in terms of R_d and strain amplitudes. For group $R_d = 0.1$, $R_d = 0.5$, and $R_d = 0.9$ (Fig. 4.14) when the total strain is smaller than 0.4%, the ascending branches of the loops are almost coincident, exhibiting nearly ideal Masing behavior. When the total strains are bigger than 0.4%, the ascending branches deviate from Masing-type behavior. This phenomenon can

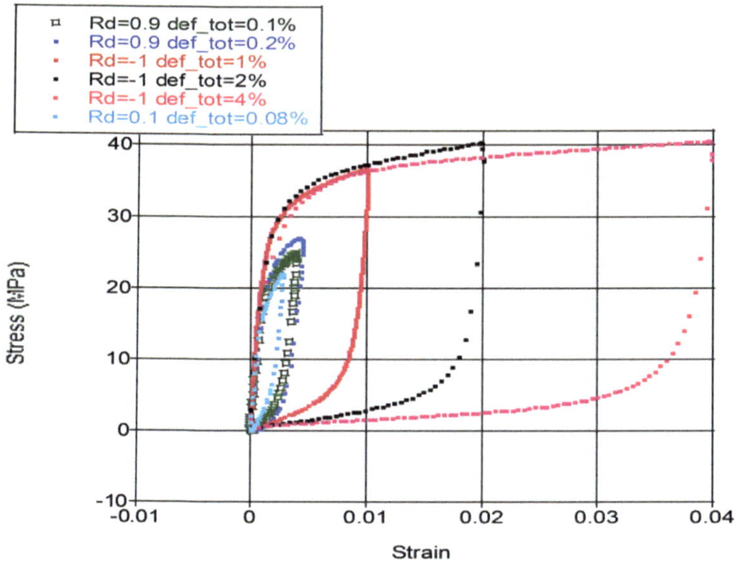

Figure 4.13 Stable hysteresis curves with different R_d values and total strain values.

be explained on the basis of the studies of Pual, Maier, and Plumtree [21–25]. From a microscopic point of view, Masing behavior relates to a stable microstructure condition and dislocation substructure [24]. When the cyclic deformation is controlled by matrix properties, non-Masing behavior should be observed [25]. For the SACNi solder, the fact that Masing behavior is observed under a small total strain means the dislocation mechanism is the same for low strain ranges. After a threshold value, probably governed by strain amplitude, the deformation mechanism changes. Even though this threshold value is not determined in this study, we can observe that the existence of the threshold is not influenced by the R_d value.

The hysteresis loops of the same total strain for different R_d values are also plotted (Fig. 4.15). According to Figs. 4.15a and 4.15b R_d has no significant effect on the loop form when the total strain is small. For larger total strains, R_d influences strongly the hysteresis loop (Fig. 4.15c). The material becomes softer when the R_d value is high and creep plays a more important role.

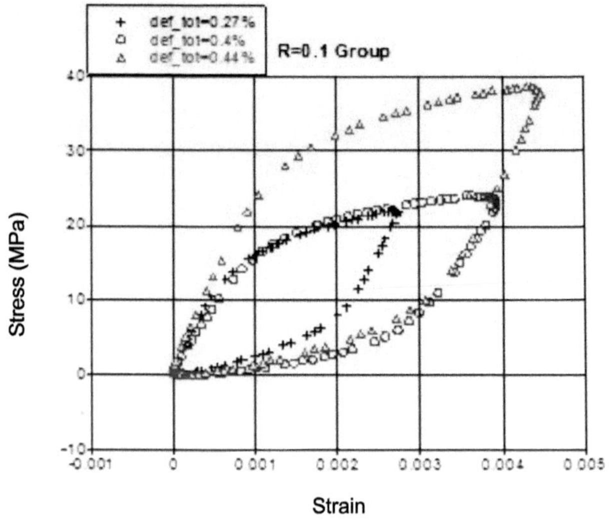

(a) Stable hysteresis curves of the $R_d = 0.1$ group

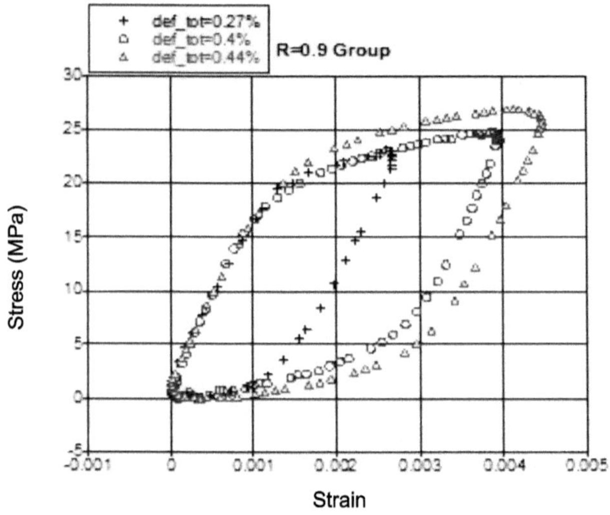

(b) Stable hysteresis curves of the $R_d = 0.9$ group

Figure 4.14 Masing behavior analysis for different R_d values.

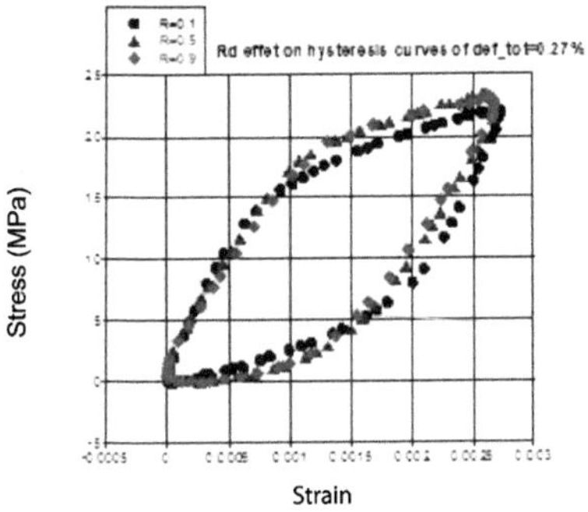

(a) Stable hysteresis curves with different R_d values (total strain 0.27%)

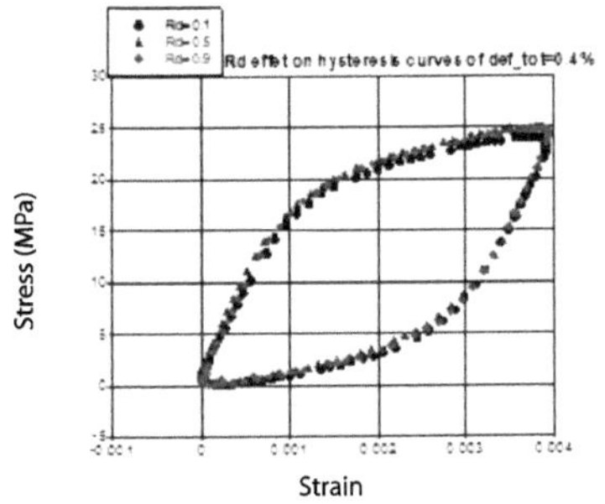

(b) Stable hysteresis curves with different R_d values (total strain 0.4%)

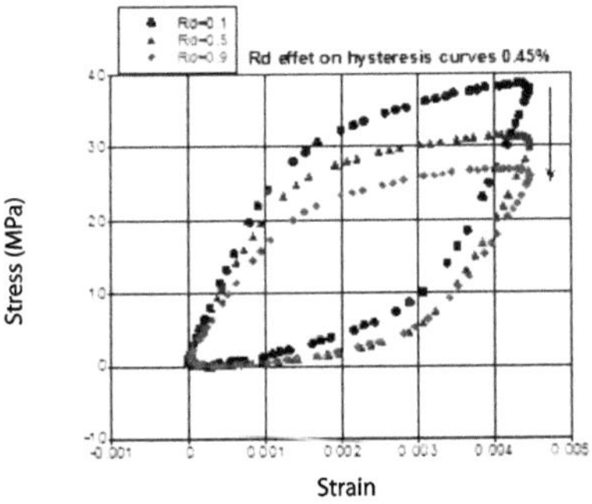

(c) Stable hysteresis curves with different R_d values (total strain 0.45%)

Figure 4.15 Analysis of stable hysteresis loops.

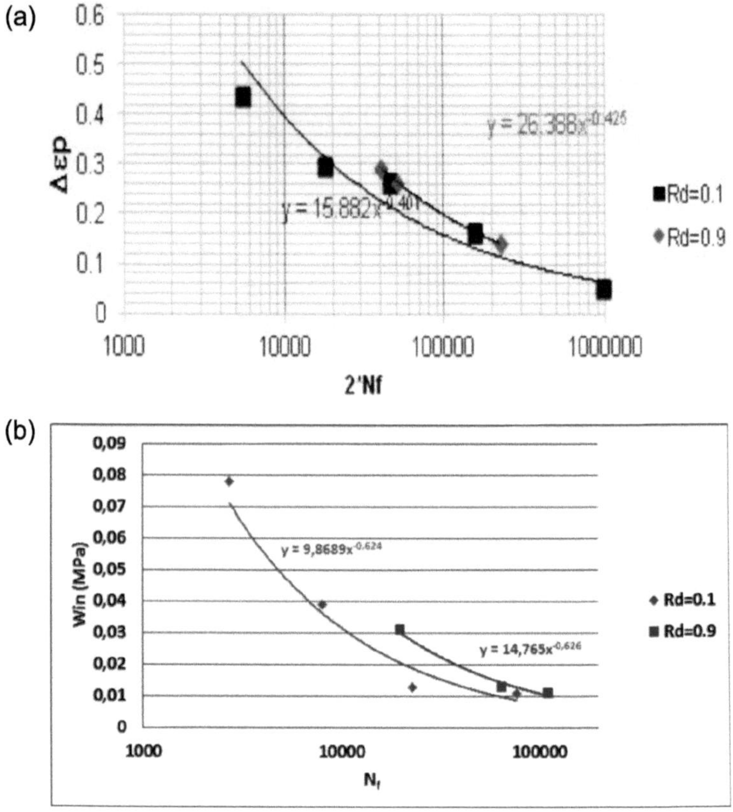

Figure 4.16 (a) Strain–life relationships with different R_d values. (b) Fatigue life in terms of stable hysteresis energy.

4.2.3 *Fatigue Life Curves and Fractography*

Strain–life curves of $R_d = 0.1$ and $R_d = 0.9$ conditions were plotted (Fig. 4.16a). Even though the strain range covered at the high R_d value is not large, it can be seen that the curves are separated when the plastic strain is large and get close to each other when the plastic strain is small. According to the strain–life curves, the material has a higher fatigue ductility coefficient when R_d increases.

In Fig. 4.16b, the same data are analyzed in terms of hysteretic energy. It follows from these figures that the effect of R_d is better illustrated by using the energy parameter [26].

Table 4.5

(a) Coffin–Manson law parameters		
ε_f	C	R_d
15.88	-0.41	0.1
26.38	-0.425	0.9
(b) Energy law parameters		
W_f	C'	R_d
9.87	-0.624	0.1
14.76	-0.626	0.9

The constants corresponding to Coffin–Manson exponents and the energy laws are given in Table 4.5.

After fatigue tests fractographic analysis is carried out. Under the same total strain, 0.44%, intergranular features are found when R_d is 0.9 (Fig. 4.17a). When creep dominates, cracks occur in the grain boundaries. Dimple fracture is also observed in this condition (Fig. 4.17b), representing ductile material behavior. Mixed features,

(a) Inter-granular cracks

(b) Dimples at final fracure

(c) Mixed fatigue features

Figure 4.17 Fatigue features analyzed by SEM.

containing striations and intergranular decohesion, were observed in the other cases (Fig. 4.17c).

4.3 Local Fatigue Behavior

In the sections described above, differences in creep behavior and moduli were observed between the bulk material and the in situ solder. Thus investigation of local fatigue behavior becomes interesting. In this study, the nanoimpact is proposed for local fatigue study. The pendulum impulse impact option [27, 28] of the NanoTest system was used for nanoimpact testing. The impact tests were carried out on a solder bulk specimen (Fig. 4.18). The probe was accelerated from a distance of 1 μm from the sample surface with an impact force of 1 mN. The computer-controlled experiments were carried out with an impact period of 6 s (3 s on and 3 s off the surface) and realized at the same location for 100 impacts. The test setup is shown in Fig. 4.19.

A typical result showing penetration depth with impact time is shown in Fig. 4.20. The impact depth increases smoothly with time without exhibiting any fracture events as can be seen in fragile materials [27, 29]. Thus the solder alloy under the impact conditions tested here exhibits ductile behavior. The impact mark is observed by scanning electron microscopy (SEM) (Fig. 4.21), and the grains

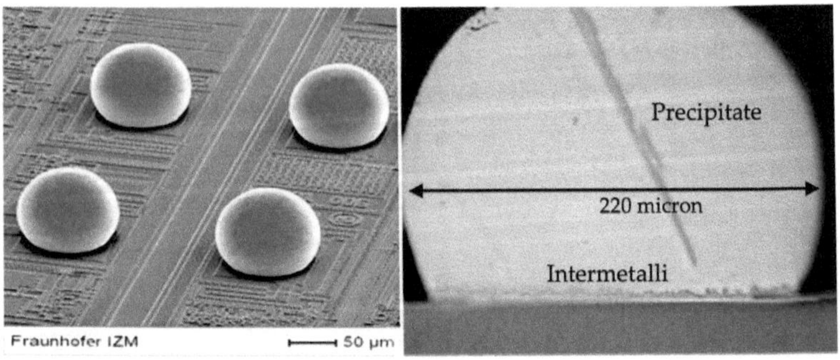

Figure 4.18 Solder connections and a microsection of a solder joint of a flip chip.

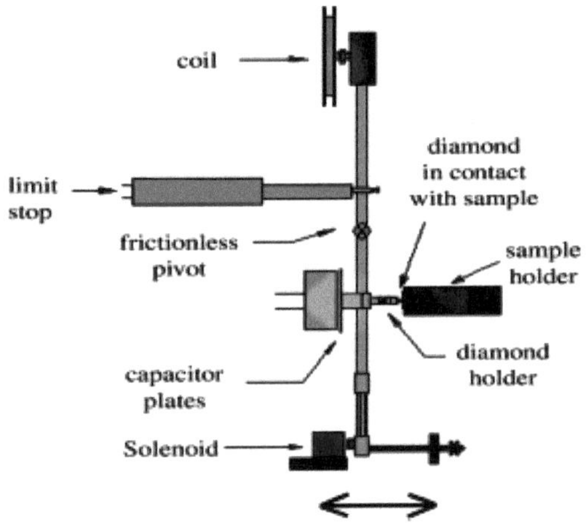

Figure 4.19 NanoTest system configuration for impact testing.

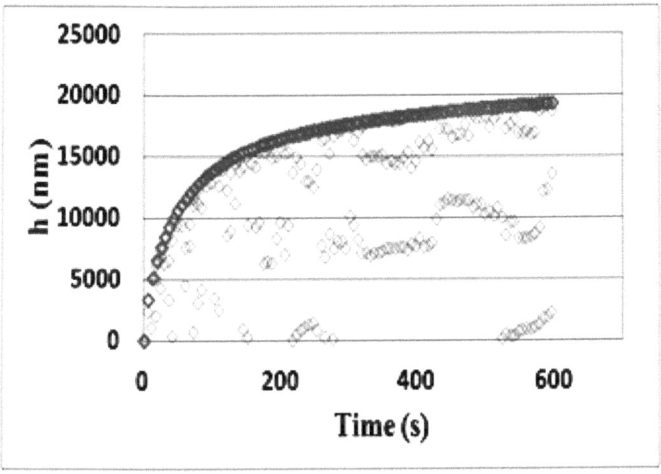

Figure 4.20 Nanoimpact test result.

boundaries are visible. This is the result of the manifestation of substructure dislocation.

Impact tests were then carried out on the insitu solder ball, one on an area clear off intermetallic precipitates and the other on the intermetallic (Fig. 4.22).

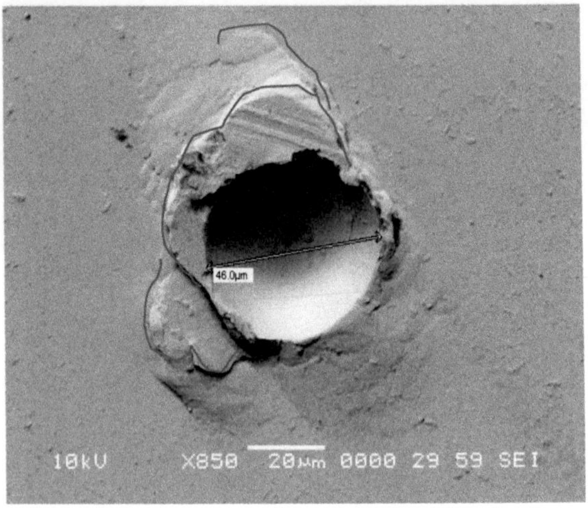

Figure 4.21 Impact mark observation by SEM.

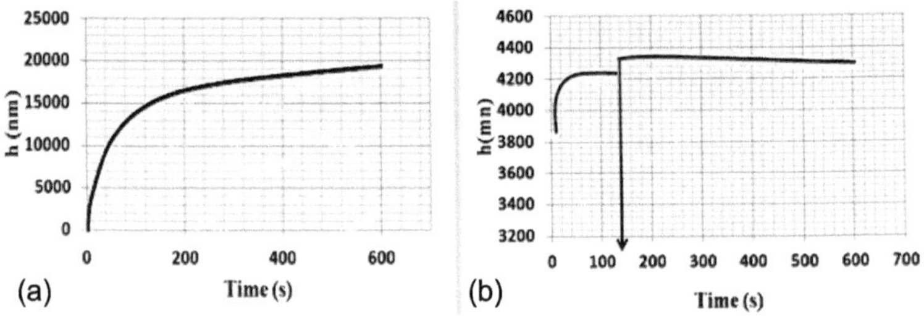

Figure 4.22 Impact depth vs. time (a) on the solder ball and (b) on the intermetallic.

It can be seen here that on the solder ball, the impact depth increases rapidly first and then almost stabilizes. On the intermetallic, we see a "break event" at about 140 s impact. This should correspond to a local failure.

The energy absorbed by the impact can be analyzed using a simple linear system [30].

$$W_N = \sum_{i=2}^{N} W_i = \sum_{i=2}^{N} \frac{1}{2} * F * \Delta h \tag{4.2}$$

Here, W_i is the energy per impact, Δh the penetration depth, and W_N the number of impacts.

This value is drawn from the impact test data, and we can determine the energy density to crack initiation by

$$Q_N = W_N / V_i \tag{4.3}$$

Here V_i is the volume of the crater created by the impact, measured by SEM.

We obtain a value of 6.8 MPa for Q_N.

From the test conditions, 140 s correspond to 28 impacts; thus this corresponds to an average of 0.28 MPa per cycle. We can now estimate the corresponding energy from the energy law defined above from bulk alloy tests (Table 4.5b). We get a value of 1.23 MPa. We can also determine the equivalent number of cycles to failure for the bulk alloy corresponding to an energy of 0.28 MPa—this give a value of 295 cycles.

This brief analysis again highlights the differences between the bulk alloy and the in situ solder.

It should be kept in mind that the local failure analysis was possible only on the intermetallic, as the failure event was clearly identified. Further tests are necessary to confirm these results

4.4 Summary and Conclusions

The mechanical properties of a lead-free solder have been studied by conventional mechanical tests and by nanoindentation. The results show that the in situ solder has a higher modulus and creep resistance than the conventional bulk solder.

The cyclic behavior of the bulk alloy has been extensively analyzed to highlight the effects of loading conditions. The material shows Masing behavior at low strain amplitudes and low displacement ratios. At high displacement ratios creep effects modify the material behavior.

Preliminary nanoimpact fatigue tests have been carried out under in situ conditions. They indicate that the energy to impact failure for the in situ solder can be lower than that for the bulk alloy, due to the presence of intermetallic precipitates.

References

1. Chong, D.Y.R., Chen, F.X., Pang, J.H.L., Ng, K., Tan, J.Y.N., and Low, P.T.H. (2006). Drop impact reliability testing for lead-free and lead-based soldered IC packages, *Microelectron. Reliab.*, **46**(7), 1160–1171.
2. Pang, J.H.L., and Xiong, B.S. (2004). Low cycle fatigue study of lead free 99.3Sn–0.7Cu solder alloy, *Int. J. Fatigue*, **26**(8), 865–872.
3. Kanchanomai, C., Miyashita, Y., and Mutoh, Y. (2002). Low-cycle fatigue behavior of Sn-Ag, Sn-Ag-Cu, and Sn-Ag-Cu-Bi lead-free solders, *J. Electron. Mater.*, **31**(5), 456–465.
4. Kariya, Y., Hosoi, T., Kimura, T., Terashima, S., and Tanaka, M. (2004). Low cycle fatigue properties of Ni added low silver content Sn-Ag-Cu flip chip interconnects, *Mater. Trans.*, **45**(3), 689–694.
5. Kariya, Y., and Suga, T. (2007). Low-cycle fatigue properties of eutectic solders at high temperatures, *Fatigue Fract. Eng. Mater. Struct.*, **30**(5), 413–419.
6. Lin, C.-K., and Huang, C.-M. (2006). Effects of strain ratio and tensile hold time on low-cycle fatigue of lead-free Sn-3.5Ag-0.5Cu solder, *J. Electron. Mater.*, **35**(2), 292–301.
7. Andersson, C., and Liu, J. (2008). Effect of corrosion on the low cycle fatigue behavior of Sn–4.0Ag–0.5Cu lead-free solder joints, *Int. J. Fatigue*, **30**(5), 917–930.
8. Kanchanomai, C., Miyashita, Y., Mutoh, Y., and Mannan, S. (2003). Influence of frequency on low cycle fatigue behavior of Pb-free solder 96.5Sn–3.5Ag, *Mater. Sci. Eng. A*, 90–98.
9. Goodall, R., and Clyne, Y.W. (2006). A critical appraisal of the extraction of creep parameters from nanoindentation data obtained at room temperature. *Acta Mater.*, **54**, 5489–5499.
10. Gao, F., and Takemeto, K. (2006). Mechanical properties evolution of SnAg3.5 based lead free solders by nanoindentation, *Mater. Lett.*, **60**, 2315–2318.
11. Liu, C.Z., and Chen, J. (2006). Nanoindentation of lead-free solders in microelectroic packaging, *Mater. Sci. Eng. A*, **448**, 340–344.
12. Gao, F., Nishikawa, H., Takemoto, T., and Qu, J. (2009). Méchanical properties versus temperature relation of individual phases in SnAg3Cu0.5 lead free solder alloy, *Microelectron. Reliab.*, **49**, 296–302.
13. Dong, K., and Shang, G. (2005). Lead free solder interconnect reliability, Dongkai Shangguan, Ed. ASM International, **77**.

14. Jasbir, B. (2007). Lead free soldering, lead free solder joint reliability, Springer **160**.
15. Ma, Z., Chalon, F., Leroy, R., Ranganathan, N., and Beake, B.D. Local and global properties of a lead free solder, *J. Electron. Mater.*, **43**, 658–670.
16. Fakpan, K., Otsuka, Y., Mutoh, Y., Inoue, S., Nagata, K., and Kodani, K. (2011). Creep-fatigue crack growth behavior of Pb-contained and Pb-free solders at room and elevated temperatures, *Proc. Eng.*, **10**, 1238–1243.
17. Paul, S.K., Sivaprasad, S., Dhar, S., and Tarafder, S. (2011). Key issues in cyclic plastic deformation: experimentation, *Mech. Mater.*, **43**(11), 705–720.
18. Feltner, C.E., and Laird, C. (1967a). Cyclic stress-strain response of FCC metals and alloys-I.phenomenological experiments. *Acta Metall.*, **15**, 1621–1632.
19. Eifler, D., and Macherauch, E. (1990). Microstructure and cyclic deformation behavior of plain carbon and low-alloyed steels, *Int. J. Fatigue*, **12**(3), 165–174.
20. Jiang, Y., and Zhang, J. (2008). Benchmark experiments and characteristic cyclic plasticity deformation, *Int. J. Plasticity*, **24**(9), 1418.
21. Paul, S.K., Sivaprasad, S., Dhar, S., and Tarafder, S. (2010a). Ratcheting and low cycle fatigue behavior of SA333 steel and their life prediction, *Int. J. Nucl. Mater.*, **401**(3–1), 17–24.
22. Paul, S.K., Sivaprasad, S., Dhar, S., and Tarafder, S. (2010c). Cyclic plastic deformation and cyclic hardening/softening behavior in 304LN stainless steel, *Theor. Appl. Fract. Mech.*, **54**(1), 63–73.
23. Kanchanomai, C., Miyashita, Y., and Mutoh, Y. (2002). Low cycle fatigue behavior of lead free solder 96.5Sn/3.5Ag, *J. Electron. Mater.*, **3**(2), 142–151.
24. Maier, H.J., Gupta, N., Karaman, I., and Haouaoui, M. (2006). Cyclic stress strain response of ultrafine grained copper, *Int. J. Fatigue*, **28**, 243–250.
25. Plumtree, A., and Abdel-Raouf, H.A. (2001). Cyclic stress strain response and substructure, *Int. J. Fatigue*, **23**, 799–805.
26. Ellyin, F., and Kujawaski, D. (1984). Plastic strain energy in fatigue failure, *J. Pressure Vessel Technol.*, **106**, 342–347.
27. Beake, B.D., and Smith, J.F. (2004). Nano-impact testing: an effective tool for assessing the resistance of advanced wear-resistant coatings to fatigue failure and delamination, *Surf. Coat. Technol.*, **188–189**, 594–598.

28. Beake, B.D., Goodes, S.R., Smith, J.F., et al. (2002). Investigating the fracture resistance and adhesion of DLC films with micro-impact testing, *Diam. Relat. Mater.*, **11**, 1606.
29. Beake, B.D., Lau, S.P., and Smith, J.F. (2004). Evaluating the fracture properties and fatigue wear of tetrahedral amorphous carbon films on silicon by nano-impact testing, *Surf. Coat. Technol.*, **177–178**, 611–615.
30. Ma, Z., Belhenini, S., Chalon, F., Leroy, R., Ranganathan, N., and Dosseul, F. (2012). Energy density estimation of crack initiation in Sn-Ag-Cu(Ni) solder by nano-impact, *La Fiabilité Mécanique des Composants Electroniques, Tours.*

Chapter 5

Determination of Plastic Zone Sizes at the Crack Tip

Tien Dung Do, Florent Chalon, René Leroy, Damien Joly, and Naryanaswami Ranganathan
Laboratoire de Mécanique et de Rhéologie (LMR) EA2640, Université François Rabelais de Tours (UFRT), Polytech 'Tours, 7 avenue Marcel Dassault, Tours, 37200, France
ranganathan@univ-tours.fr

5.1 Introduction

Fatigue crack initiation is controlled by cyclic stress and yield strength of the material. Fatigue cracks initiate at positions where the cyclic plastic deformation is higher than average. Generally, the cracks initiate either at or near the specimen surface [1]. There is no generally accepted quantitative description of the crack initiation process, and consequently, it is quite difficult to perform a numerical analysis on the crack initiation [2, 3].

The plastic zone (PZ) at the crack tip plays an important role in determining crack initiation and crack propagation kinetics [4–7]. Therefore, it is important to know the exact size and shape of the PZ at the fatigue crack tip and how they change under loading. The high stresses around the crack tip exceed the yield strength of materials

Materials Characterization: Modern Methods and Applications
Edited by Narayanaswami Ranganathan

ISBN 978-981-4613-06-4 (Hardcover), 978-981-4613-07-1 (eBook)
www.panstanford.com

and induce a small plastic deformation volume called the *plastic zone (PZ)* at the crack tip. This PZ shields the crack tip from high stresses. If the PZ is large, a high amount of energy is dissipated during crack propagation and when the PZ is small, crack propagation requires less energy.

The mechanism of fatigue crack initiation has been investigated for a long time, and many valuable conclusions have been published [8–10]. However, the mechanisms of fatigue crack initiation are still a matter of speculation. The shape and size of the PZ at the crack tip and quantitative prediction of the location of fatigue crack initiation have not been systemically analyzed and reported until now.

Theoretical models exist to predict the size and shape of the PZ. These models, due to Irwin, Rice, Dugdale, and other plasticity models, are discussed by Jendhoubi [11]. Analysis of these models shows that there is no consensus on the relationships between the size of the PZ and the fracture mechanics parameters.

There are many methods [12] to study the PZ at the crack tip, such as microhardness measurements, the etching technique, the optical interference technique, and the X-ray microbeam (XMB) technique. The nanoindentation technique has been an increasingly popular technique for material characterization in the nanoscale. Developed in the early 1970s [13], nanoindenters have since been developed, and this technology is widely available for researchers interested in thin films and small volumes as motivated by modern applications. Nanoindentation was originally developed to measure the hardness and elastic properties of the material of interest.

This chapter presents the current techniques for estimating and measuring the fatigue PZ at the crack tip. The nanoindentation technique and optical profiling system were used to study the PZ for aluminum alloy 2024T351.

5.2 Definition of the Fatigue Plastic Zone at the Crack Tip

During the load increase from 1 to 2 (Fig. 5.1a), the material at the crack tip will plasticize when the local stresses reaches the yield strength. The behavior of material (assuming an elastic–

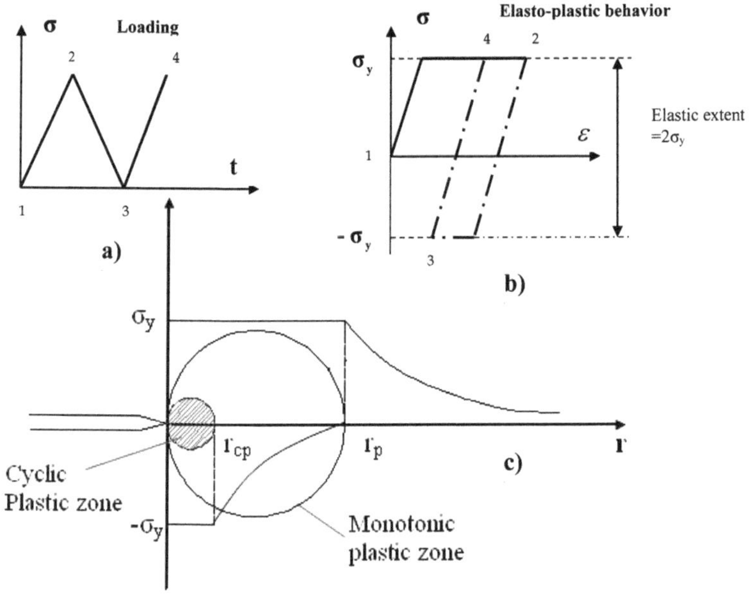

Figure 5.1 Schematic illustration of the monotonic and cyclic plastic zone at the crack tip.

plastic perfect material) in this process is shown in Fig. 5.1b. The process induces a plastic deformation, as in Fig. 5.1c, called the monotonic PZ, where the stresses are leveled off at the yield stress of the material. The theoretical PZ size prediction at the crack tip under mode I loading is of the form given in Eq. 5.1 [11]. The maximum monotonic PZ size r_p for all materials correlates linearly with $(K_{max}/\sigma_y)^2$, where K_{max} is the maximum stress intensity factor during a fatigue cycle and σ_y is the yield strength of the material. For most materials, the correlating coefficient α lies between 0.05 and 0.14. Notable exceptions are the values for aluminum alloys 2024-T3, 2024-T4, and 7075-T6 with, respectively, α equal to 0.3, 0.25, and 0.4 [13].

$$r_p = \alpha \left(\frac{K_{max}}{\sigma_y} \right)^2 \tag{5.1}$$

In a fatigue cycle, when the load decreases from 2 to 3 (Fig. 5.1a), the material induces a plastic deformation in compression at the crack tip. The plastic deformation in compression occurs when the compressive stress amplitude at the crack tip is equal to twice the yield strength (see Fig. 5.1b). The appearance of the cyclic plastic deformation zone is called the cyclic PZ (see Fig. 5.1c). The radius of this area is given by a relation similar to Eq. 5.1 by replacing K by ΔK and σ_y by $2\sigma_y$, ΔK being the stress intensity factor amplitude. Hence

$$r_{cp} = \alpha \left(\frac{\Delta K}{2\sigma_y}\right)^2 \tag{5.2}$$

This cyclic PZ is four times smaller than the monotonic PZ in the case of the load ratio $R = 0$. In general, the relationship between the cyclic and the monotonic PZ is given by [14]

$$\rho_{(\text{cyclic/monotone})} = \frac{(1-R)^2}{4} \tag{5.3}$$

where R is the load ratio.

We shall now briefly discuss current techniques for measuring PZ size and shape.

5.3 Current Techniques for Estimating and Measuring the Fatigue Plastic Zone at the Crack Tip

5.3.1 *Microhardness Measurement*

The microhardness technique is applicable to materials that strongly work-harden (like austenitic stainless steels) or work-soften (like maraging steels). The procedure is simply taking hardness indentations ahead or around the crack tip and determining the elastic–plastic boundary as a sudden variation in hardness. Bathias and Pelloux [15] used this technique to determine the monotonic and cyclic plane strain PZ sizes of maraging and austenitic stainless steels as a function of ΔK. They used 25 g and 50 g indentation loads, and the size of the indentation was 10 μm. They found the cyclic PZ sizes to be one-fourth of the monotonic PZ sizes. They also argue that a local work hardening due to plastic strains as small as 1% can be detected by microhardness measurements.

The microhardness technique was employed in a detailed study by Saxena and Antolovich [16] on fatigue crack propagations of four

Cu–Al alloys. It has also been applied on aluminum alloys by Bathias [17] and Ranganathan and Petit [18] and on a single crystal of pure copper by Purcell and Weertman [19].

5.3.2 *Etching of the Plastic Zone*

In this technique, sections of the specimens containing the crack are etched with an appropriate etchant. Because the deformed and undeformed regions of the material respond differently to etching, the size and shape of the PZ can be easily observed as a dark- or light-etched region.

Hahn and Rosenfield [20] and Hahn et al. [21] applied the technique to reveal the PZ in plates of Fe-3 wt.% Si. Specimens having different thicknesses were fatigued to different numbers of cycles, and ΔK values varied from 21 to 35 MPa$\sqrt{m}$. After the fatigue cracking, the specimens were first aged to decorate the dislocations; then the sections were polished and etched. They observed a highly strained, light-etching region and a lightly strained, dark-etching region. They argued that the cyclic zone corresponds to the lightly etched region whose size was in fact one-fourth that of the monotonic PZ.

Bathias and Pelloux [15] used the etching and microhardness techniques together. Although they found that it was time consuming, they could successfully conduct the measurements by using the etching method and found PZ sizes equivalent to those measured by the microhardness technique.

5.3.3 *Optical Interference Technique*

In interference microscopy, a parallel beam of monochromatic light normal to the surface of the specimen is used. An optical flat with high reflectivity is placed in contact with the specimen and slightly tilted to produce a thin wedge between the two. The light is repeatedly reflected between the specimen and the reference flat, and interference between the two light beams takes place. However, to obtain sharp fringes, the reflectivity of the specimen surface should be the same as that of the reference plate. If a variation such as a step is present on the surface of the specimen, the light path

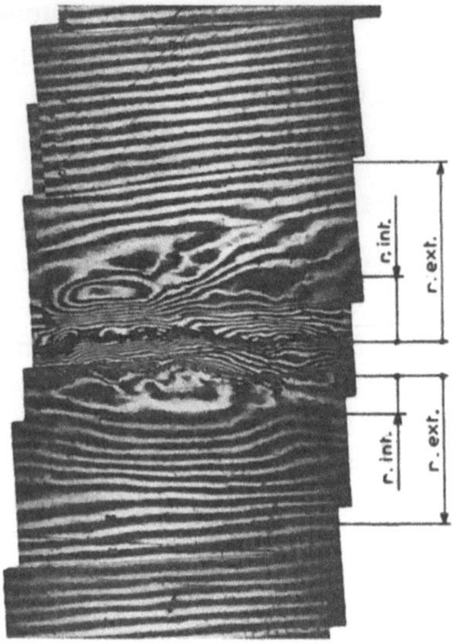

Figure 5.2 Interference fringes near the crack [22].

is disturbed and the interference fringe is displaced one way or the other. A total fringe displacement corresponds to a change in height of half the wavelength of the monochromatic light used. In material deformation studies, the interference of the light with the slip line steps in the PZ (in plane stress) makes it possible to determine the elastic–plastic boundary. A slightly different technique has been used by Ranganathan [22], where he used the Nomarski interference technique in a metallurgical microscope. An example is given in Fig. 5.2, where two PZs near the crack are observed, an interior zone associated with the cyclic PZ and an exterior zone associated with the monotonic PZ.

5.3.4 *Foil Strain-Gauge Technique*

In the foil strain-gauge technique, a tiny strain-gauge is situated on the specimen so that it is parallel to the loading direction ahead of the crack tip. During loading, the output from the strain-gauge is recorded as a stress–strain diagram. When strain becomes nonlinear, the distance from the crack tip to the strain-gauge center is determined as the PZ size. Being a surface measurement, this method is only applicable to the plane stress PZ.

Izumi and Fine [23] and Jendhoubi [11] applied this method on a series of Al alloys and on a Ni–Al alloy and measured the PZs as a function of ΔK. They found a monotonic PZ that was two times as large as the cyclic PZ. They also mapped the strain distributions within the "butterfly wing shaped" PZs.

5.3.5 *Recrystallization Method*

It is well known that the recrystallization temperature depends on the amount of stored energy of deformation in a material and the recrystallized grain size is a function of the amount of this energy and of the annealing temperature [24]. The deformed region around a crack or at a crack tip can therefore be revealed as a large grained region by annealing the material at appropriate temperatures. Tschegg et al. [25] measured PZs around fatigue cracks in mild steel with the help of recrystallization and XMB techniques. They enumerated some difficulties and limitations of the recrystallization technique regarding strain resolution and different dislocation structures in the cyclic and monotonic PZs.

Iino [24] studied the recrystallization of deformed grains in 304 stainless steel and stated that recrystallization occurred when this material was strained $>2\%$.

Kwun and Park [26] also applied the recrystallization method on notched α-brass plates. They emphasized that the advantages of this method are that one can see the real shape and size of the PZ and one may also deduce the plastic strain distribution within the zone.

5.3.6 *X-Ray Microbeam Technique*

An X-ray diffraction pattern of an undeformed crystal appears as a series of sharp spots. However, if the material is strained, owing to a change in diffraction conditions, Debye–Scherrer (DS) rings replace the sharp single reflections.

Tschegg et al. [25] employed the XMB technique as well as the recrystallization method to measure PZs around fatigue cracks in mild steel. They took a number of diffraction patterns along the crack and perpendicular to the crack edge, and by measuring the DS line broadening, they mapped the PZ around the fatigue cracks.

5.3.7 *Shear Lip Size Measurements*

Lai and Ferguson [27] measured the shear lip size of steel and 7075-T6 aluminum alloy compact tension specimens after they had been broken apart. They found the following relation between the plane strain fracture toughness and the size of the shear lip (B_{SL}):

$$B_{SL} = 0.41 \left(\frac{K_{IC}}{\sigma_{ys}} \right)^2 \tag{5.4}$$

They argue that this equation is similar to Dugdale's theoretical plane stress PZ. Therefore, $B_{SL} = r_p$. The shear lip size is not only a representation of the PZ but also a measure of its size. However, only the ultimate PZ size corresponding to fracture (K vales approaching K_c, the fracture toughness) were measured.

Hertzberg [28] also correlates the size of the PZ to the shear lip size found on the fracture surface of a material.

5.3.8 *Selected-Area Channeling Pattern Technique*

Electron channeling can be briefly explained as a mechanism of contrast obtained when either back-scattered or forward-scattered electrons, with approximately the same energy as those in the beam, are used in an imaging mode in the scanning electron microscope (SEM). When an impinging electron beam interacts with the planes of an illuminated crystal, electron diffraction occurs. When this beam is swept across the crystal, diffraction conditions change with the change of the incidence angle. As a result, a diffraction pattern may be observed [29]. Among many theoretical treatments of this diffraction process, that of Spencer and Humphreys is the most quantitative [30].

Channeling patterns may be produced from a very small area in the form of selected-area electron channeling patterns (SACPs). In the SACP technique, the electron beam is made to rock about a spot on the specimen surface. This means physically that the scanning electron beam is directed so as to meet the electron optical axis of the microscope at the specimen surface. The size of the spot on the specimen surface from which the SACP is produced can be varied (over the range 10–1000 μm in diameter) by changing the electron optical conditions in the microscope. This makes the technique very advantageous for material deformation studies because it allows averaging of information over an area of chosen dimensions. The principles of channeling and the theory and applications have been reviewed by Schulson [31], Joy et al. [32], and Davidson [29].

Detailed information of channeling patterns in deformation studies is given in Ref. [33]. Tekin and Martin [34] illustrate the degradation of SACP with strain, the means of obtaining the best channeling patterns, and the procedure for PZ size mapping; the advantages and disadvantages of the technique are summarized elsewhere [29, 34].

By using this method, one can map the PZ at the crack tip and around the crack of any crystalline material exhibiting clear channeling patterns both in plane stress, from observations on the external surface, and in plane strain conditions, by preparing sections taken through the specimen thickness [34].

5.3.9 *Slip Lines Method*

This method essentially depends on the observation of slip lines around and at the crack tip under an optical microscope [18, 22]. It is only applicable to plane stress PZs.

5.4 Measuring the Plastic Zone Size at the Fatigue Crack Tip by Nanoindentation Technique

5.4.1 *Introduction to the NanoTest System*

Nanoindentation has become a commonplace tool for the measurement of mechanical properties at small scales but may have even greater importance as a technique for experimental studies of fundamental materials physics. With high-resolution load–displacement data, discrete events, including dislocation source

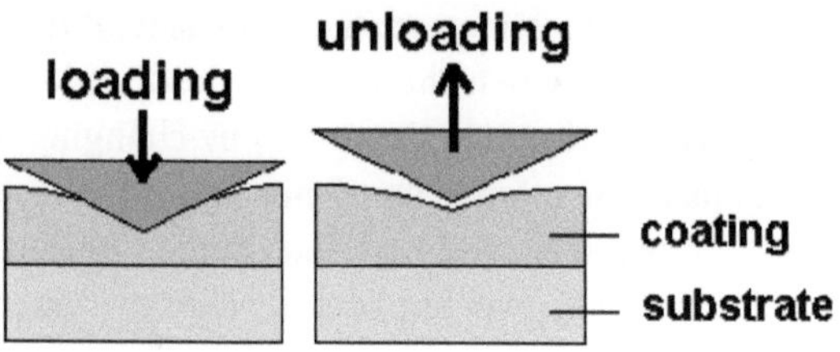

Figure 5.3 The movement of a probe in contact with a surface.

activation, shear instability initiation, and phase transformations, can be detected during a nanoindentation test. Details are given in Chapter 1 of this book.

The NanoTest system is a nanomechanical property-testing center. The system has up to three separate modules (indentation, scanning, and impact), depending on the version. All three modules can be used in conjunction with the low load head (0.1–500 mN) or the high load MicroTest head (0.1–20 N).

The NanoTest system measures the movement of a diamond probe in contact with a surface. To make a measurement an increasing load is applied and the probe indents the surface, as shown schematically below. The elastic and plastic properties of the sample are determined, thanks to the curve of unloading. The applied load, displacement, reduced modulus, hardness, and elastic and plastic energies are recorded during the indentation phase. An example on a coating material over a substrate is given in Fig. 5.3.

During a typical nanoindentation test, force and displacement are recorded as the indenter tip is pressed into the test material's surface with a prescribed loading and unloading profile. The response of interest is the load–displacement curve (often called the P–h curve), such as that depicted in Fig. 5.4 for an indentation. The global shape of the P–h curve differs from one material to the next, and these variations usually reflect different mechanical properties. In an indentation, the elastic and plastic properties of the sample determine the shape of the unloading curve. In Fig. 5.4, the different characteristics that can be obtained are depicted.

Figure 5.5 shows a typical set of results and a screen shot of the relevant measured parameters.

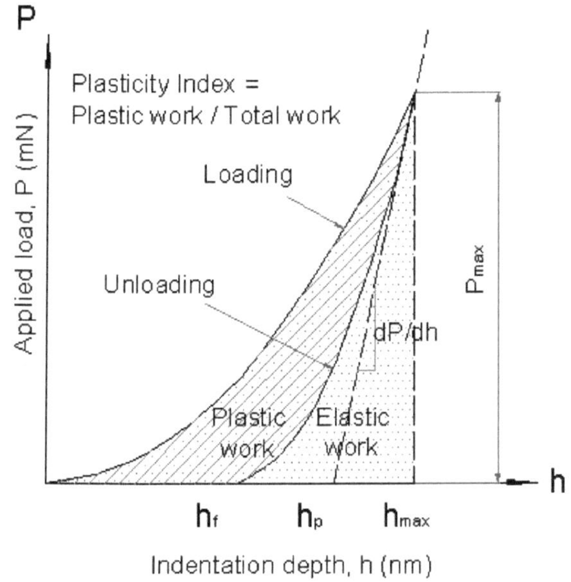

Figure 5.4 Example of a typical load–displacement curve (P–h curve).

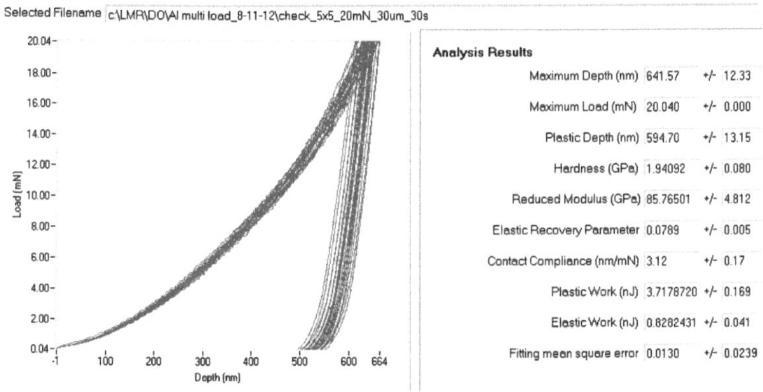

Figure 5.5 Example of results of load–displacement curves.

In the current study, we use this technique to identify the plastically deformed zone near a fatigue crack in an aluminum alloy. The experimental details are now described below.

Table 5.1 Composition (in wt.%) of the studied aluminum alloy (Al not indicated)

Alloy	Si	Fe	Cu	Mn	Mg	Cr	Zn	Ti
2024T351	0.1	0.22	4.46	0.66	1.5	0.01	0.04	0.02

Table 5.2 Nominal mechanical properties

Alloy	Young's modulus (GPa)	Yield strength (MPa)	UTS (MPa)	Elongation (%)
2024T351	73	300	500	16

5.4.2 *Fatigue Crack Growth Test*

We used the aluminum alloy 2024T351 in this study. The nominal chemical composition and mechanical properties are shown in Tables 5.1 and 5.2.

Fatigue crack growth tests were conducted on a compact tension (CT) specimen with 4.5 mm thickness. Figure 5.6 illustrates the main dimensions of the sample used in this chapter.

The notch of the specimen was machined with a wire cutter and the surface of specimen was polished up to 1 μm. The test was carried out at room temperature by using Instron tensile test

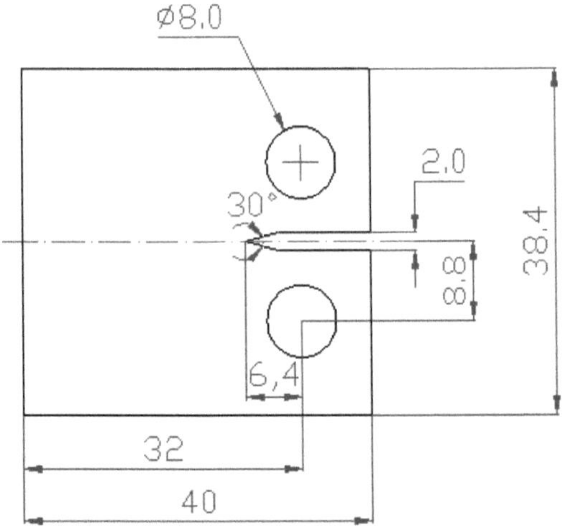

Figure 5.6 Dimensions of the compact tension specimen in this study.

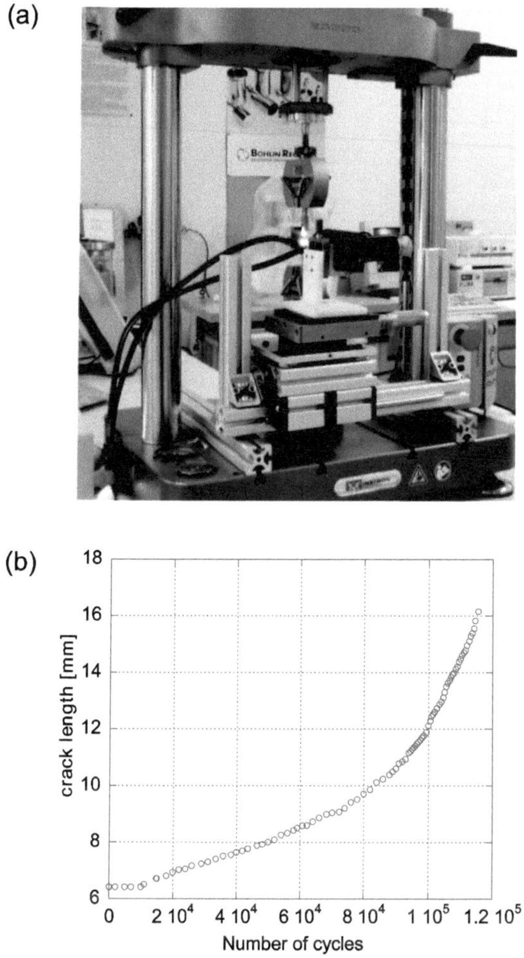

Figure 5.7 (a) Fatigue test on machine Instron E3000. (b) Crack growth curve.

machine E3000 with cyclic fatigue loads between 150 N and 1500 N, a frequency of 5 Hz, and a sinusoidal signal. The fatigue crack length was measured by using an optical microscope device. The stress intensity range ΔK increased from 7 MPa$\sqrt{\text{m}}$ to 16 MPa$\sqrt{\text{m}}$ when the fatigue crack length changed from 6.4 mm to 16 mm. The testing machine E3000 is shown in Fig. 5.7a and the crack growth curve is

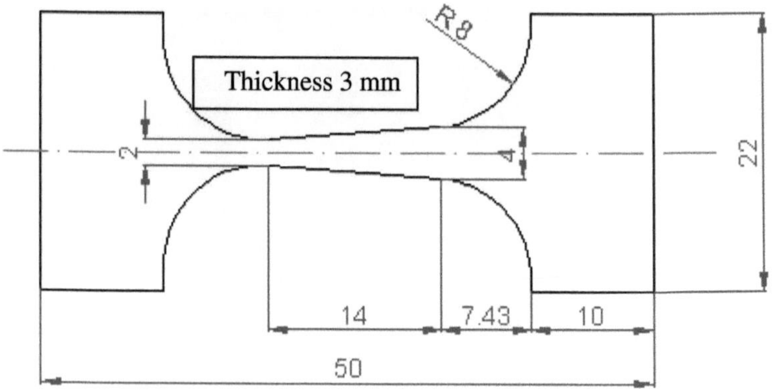

Figure 5.8 Trapezoidal test specimen.

shown in Fig. 5.7b. In another series of tests the effect of constant ΔK tests at 16 MPa.m$^{1/2}$ was studied at load ratios of 0.1 and 0.3.

Finally, with a view to determine the equivalent strains in the PZ, a calibration test was carried out in a trapezoidal test specimen (Fig. 5.8).

Following a tensile test, strains are varying along the specimen length and nanoindentaion tests permit correlate strain with relevant hardness values.

5.4.3 *NanoTest*

The indentations were performed with an ultralow load by using a Berkovich pyramid tip. This instrument is controlled by a computer; both load and displacement were recorded continuously during the whole indentation cycle. Each indentation is made in an exact manner by setting the experimental parameters. In this study we used a maximum applied load of 20 mN with an initial load of 0.08 mN, loading and unloading times of 30 s, and a dwell period at a maximum load of 60 s. The indentations were performed at the fatigue crack tip with $\Delta K = 16$ MPa $\cdot$ m$^{1/2}$ in five directions: 0°, 45° on the left, right, and 90° on the left, right, as shown in Fig. 5.9. At each direction, 80 indentations were made and each imprint was separated by 20 μm from its nearest neighbor. The exact position of the first indentation at the fatigue crack tip is observed

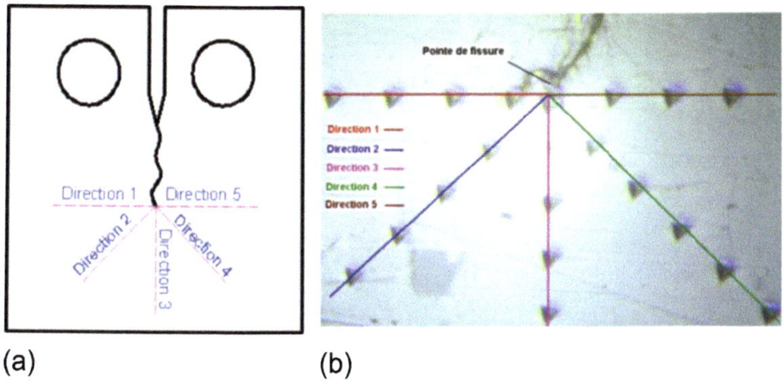

Figure 5.9 (a) The directions for nanoindents. (b) SEM picture of indentation.

by a microscope attached on the NanoTest machine. After making the indentation for five directions, the position, size, and shape of each indentation were observed in an SEM to remove all indentation that had faulty shape and size. After analysis, each indentation gives values of hardness, reduced modulus, plastic work, elastic work, and plasticity index. The measurements were made along five directions with respect to the mean crack plane (Fig. 5.9a). Figure 5.9b shows an SEM picture of the indents.

5.4.4 *Experimental Results and Analysis*

In this chapter, the PZ size at the fatigue crack tip is determined, thanks to the distribution of hardness near the fatigue crack tip in the PZ.

Figure 5.10a shows the results for direction 3.

Starting from the crack tip, where the hardness value is the highest, there is a sharp decrease leading to the beginning of almost a plateau where the hardness is almost constant. The beginning of this plateau is assimilated to the cyclic PZ. After the plateau there is a small decrease in hardness. We associate this second change in hardness with distance as the frontier of the monotonic PZ. These points are materialized with two indicators. It should be noted that the hardness values are relative with respect to the plateau value, which is taken to be 1.

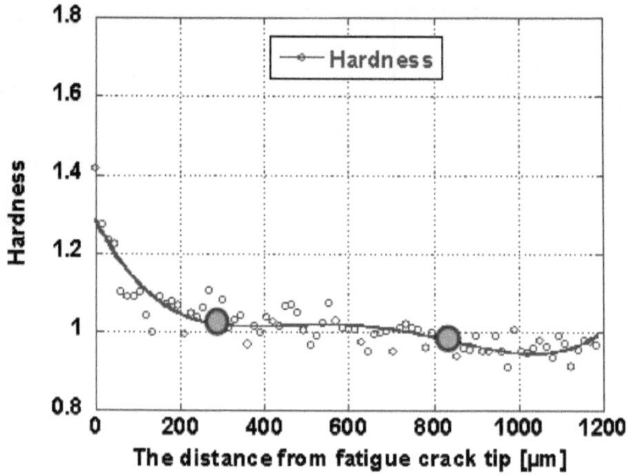

Figure 5.10 Hardness vs. distance from crack tip for direction 3.

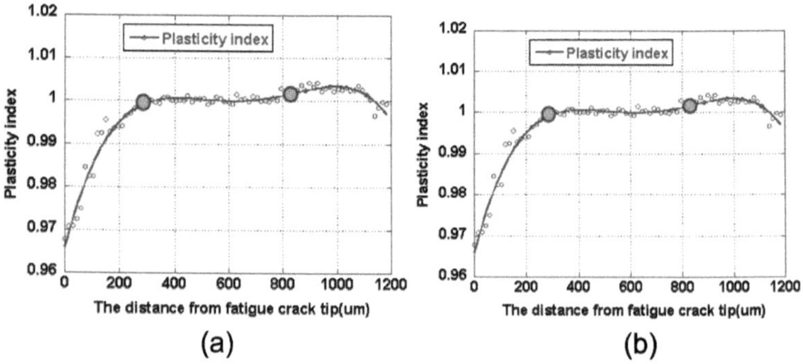

Figure 5.11 (a) Plastic work evolution near the crack. (b) Plasticity index near the crack.

The next figures, Figs. 5.11a and 5.11b, show the evolutions of normalized values of plastic work and plasticity index, respectively.

According to Fig. 5.11a, it can be clearly seen that the plastic work decreases drastically by about 15% with respect to the far-field value. The same evolution can be seen in terms of the plasticity index (Fig. 5.11).

These results show that the capacity of plastic deformation and plastic work decreases as we approach the crack tip, depending on work hardening. This adds proof to the ductility exhaustion model at the crack tip.

Table 5.3 Plastic zone sizes in the five directions compared to theoretical estimates

Directions	Cyclic plastic zone size (μm)	Monotonic plastic zone size (μm)
Direction 1	220	640
Direction 2	220	715
Direction 3	285	810
Direction 4	200	720
Direction 5	200	620
Direction 1 and 3 theory	167	571

Table 5.4 Plastic zone sizes for different load ratios

	Theory		Nanoindentation	
Load ratio (R)	Cyclic PZ (μm)	Monotonic PZ (μm)	Cyclic PZ (μm)	Monotonic PZ (μm)
0.1	166,7	571,7	220	640
0.3	166,7	923	180	820
0.5	166,7	1811	200	1750

The following table (Table 5.3) compares the cyclic and monotonic PZ sizes measured by the two techniques and the theoretical estimates, using Eqs. 5.1 and 5.2 with an α value of $1/2\pi$.

It can be seen here that the cyclic PZs measured by both the measured cyclic PZ sizes seem to be higher than the theoretical estimates. The monotonic PZ sizes are of the same order as the theoretical ones. The PZ shape is almost circular.

In Table 5.4, we compare the measured PZ sizes at three different load ratios for the same ΔK. The sizes were estimated using the method described above.

It can be seen here that the nanoindentation technique used gives coherent results with respect to the effect of load ratio on the size of the PZ (cyclic or monotonic).

5.5 Estimation of Strain in the Plastic Zone

On the trapezoidal specimen shown in Fig. 5.8, a tensile test was carried out carefully to enter plastic behavior in the smallest section. The specimen was then unloaded. Thus we have, along the length of the specimen, varying strains.

The stress at a particular point is given by

$$S = \frac{S_{max}}{(4 + 2/x) \times 14} \quad (5.5)$$

Here S_{max} is the maximum stress (in the smallest section) and x the distance a particular point along the length from the smallest section.

A calibration test by nanoindentation is now carried out using the same conditions as for the nanotests in the PZ.

The stress–strain law was also identified by a tensile test. The following constants are obtained for the Rmaberg–Osgood relationship.

$$\varepsilon = \sigma/E + (\sigma/K)^{1/n} \quad (5.6)$$

with $E = 73.7$ GPa, $K = 544.8$ MPa, and $n = 0.048$

Table 5.5 Hardness values

Distance (mm)	Stress (MPa)	Elastic strain	Plastic strain	Total strain	Hardness (GPa)
0	477.00	6.47E-03	6.27E-02	6.92E-02	4.13
1.2	439.34	5.96E-03	1.13E-02	1.73E-02	2.95
1.4	433.64	5.88E-03	8.61E-03	1.45E-02	2.89
1.8	422.66	5.73E-03	5.05E-03	1.08E-02	2.78
2.8	397.50	5.39E-03	1.41E-03	6.80E-03	2.78
3.2	388.26	5.27E-03	8.61E-04	6.13E-03	2.72
3.4	383.79	5.21E-03	6.76E-04	5.88E-03	2.72
4	371.00	5.03E-03	3.34E-04	5.37E-03	2.69
4.6	359.03	4.87E-03	1.69E-04	5.04E-03	2.68
5	351.47	4.77E-03	1.08E-04	4.88E-03	2.66
5.6	340.71	4.62E-03	5.66E-05	4.68E-03	2.63
6.4	327.35	4.44E-03	2.46E-05	4.47E-03	2.60
7.2	315.00	4.27E-03	1.10E-05	4.28E-03	2.56
7.6	309.17	4.19E-03	7.48E-06	4.20E-03	2.56
8	303.55	4.12E-03	5.10E-06	4.12E-03	2.52
8.2	300.81	4.08E-03	4.23E-06	4.09E-03	2.52
8.8	292.89	3.97E-03	2.43E-06	3.98E-03	2.48
9	290.35	3.94E-03	2.02E-06	3.94E-03	2.46
9.4	285.38	3.87E-03	1.41E-06	3.87E-03	2.46
9.6	282.97	3.84E-03	1.18E-06	3.84E-03	2.45
11.6	260.86	3.54E-03	2.17E-07	3.54E-03	2.36
11.8	258.84	3.51E-03	1.85E-07	3.51E-03	2.35
13.4	243.72	3.31E-03	5.27E-08	3.31E-03	2.31

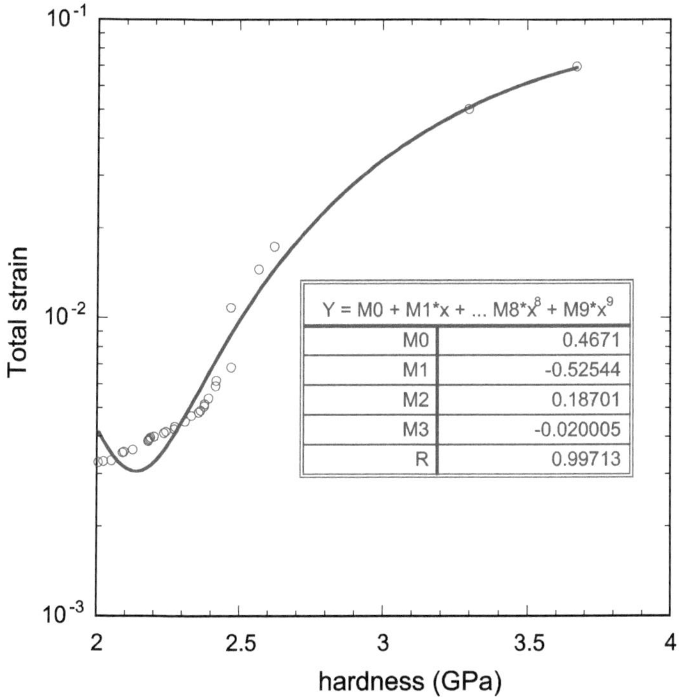

Figure 5.12 Hardness vs. total strain relationship established from a trapezoidal test specimen.

The following table (Table 5.5) shows the results correlating measured hardness to the local stresses and strains in the trapezoidal specimen.

The results for the other parameters such as plastic work and plasticity index are available in the work of Do [35].

The following figure (Fig. 5.12) shows the relationship between hardness and total strain.

The results in Fig. 5.10 can now be translated as strains in the PZ (Fig. 5.13). In this figure, equivalent strains estimated from the current analysis are compared to those theoretically estimated by Ranganathan [22]. It can be seen here that the current estimations are slightly higher than the theoretical estimations despite the experimental scatter. The estimations in Ref. [22] were carried out under plane strain conditions, while the current measurements

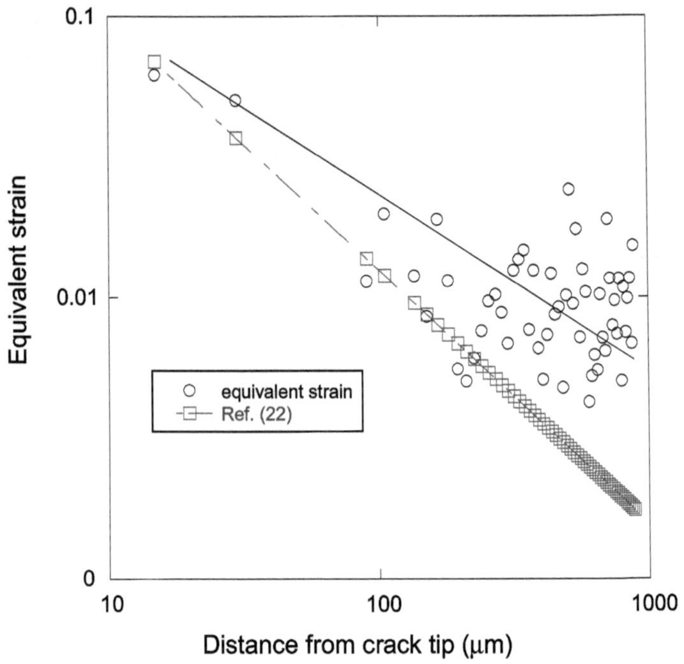

Figure 5.13 Equivalent strains in the plastic zone.

are in plane stress, on the specimen surface. This can explain the differences observed.

The current study highlights possibilities offered by the nanoindentation technique to study the deformation behavior in the vicinity of a fatigue crack.

5.6 Conclusions

Some conclusions are drawn from the present study:

(1) There are two PZs at the fatigue crack tip.
(2) Irwin's model can be used to predict the theoretical monotonic PZ size at the fatigue crack tip for the aluminum alloy 2024T351.
(3) The forms of cyclic and monotonic PZs are similar, but the sizes are different.

(4) The nanoindentation technique can measure the cyclic and monotonic PZ sizes.
(5) Strain estimations can be made in the PZ by a suitable calibration.

References

1. Xi, L. (2010). Investigation of the region of fatigue crack initiation in a transmission gear, *Mater. Sci. Eng. A*, **527**, 1377.
2. Basinski, Z.S., and Basinski, S.J. (1985). Low amplitude fatigue of copper single crystals—II. Surface observations, *Acta Metall.*, **33**, 1307–1317.
3. Cahn, R.W., and Haasen, P. (1996). *Physical Metallurgy* (Elsevier Science, Oxford).
4. Khan, S.M.A., and Khraisheh, M.K. (2000). Analysis of mixed mode crack initiation angles under various loading conditions, *Eng. Fract. Mech.*, **67**, 397e419.
5. Khan, S.M.A., and Khraisheh, M.K. (2004). A new criterion for mixed mode fracture initiation based on the crack tip plastic core region, *Int. J. Plasticity*, **20**, 55e84.
6. Li, C.B., and Kwang, S.K. (2004). The minimum plastic zone radius criterion for crack initiation direction applied to surface cracks and through-cracks under mixed mode loading, *Int. J. Fatigue*, **26**, 1169e1178.
7. Golos, K., and Wasiluk, B. (2000). Role of plastic zone in crack growth direction criterion under mixed mode loading, *Int. J. Fract.*, **102**, 341e353.
8. Xu, S., et al. (2008). Crack initiation mechanisms for low cycle fatigue of type 316Ti stainless steel in high temperature water, *Mater. Sci. Eng. A*, **490**, 16–25.
9. Yang, F., et al. (2008). Crack initiation mechanism of extruded AZ31 magnesium alloy in the very high cycle fatigue regime, *Mater. Sci. Eng. A*, **491**, 131–136.
10. Shiozawa, K., et al. (2006). Subsurface crack initiation and propagation mechanism in high-strength steel in a very high cycle fatigue regime, *Int. J. Fatigue*, **28**, 1521–1532.
11. Jendhoubi, K. (1987). Mesures locales et globales globales des paramètres mécaniques gouvernant la fissuration par fatigue (Thèse, Université de Poitiers, France).

12. Uguz, A., and Martin, J.W. (1996). Plastic zone size measurement techniques for metallic materials, *Mater. Charact.*, **37**, 105–118.
13. Poon, B., et al. (2008). An analysis of nanoindentation in elasto-plastic solids, *Int. J. Solids Struct.*, **45**, 6399–6415.
14. Henaff, G., et al. (2005). Fatigue des structures: endurance, critères de dimmensionnement, propagation des fissures, rupture (Ellipses, France).
15. Bathias, C., and Pelloux, R.M. (1973). Fatigue crack propagation in martensitic and austenitic steels, *Metall. Trans.*, **4**, 1265–1273.
16. Saxena, A., and Antolovich, S.D. (1975). Low cycle fatigue, fatigue crack propagation, and substructures in a series of polycrystalline Cu-Al alloys, *Metall. Trans. A*, **6**, 1809–1828.
17. Bathias, C. (1977). Plastic zone formation and fatigue crack grotitk, *Proc. 4th Int. Conf. Fracture II (ICF4)*, 1307–1312.
18. Ranganathan, N., and Petit, J. (1983). Quantitative measurements in the plastic zone created by an overload in air and vacuum, fatigue mechanisms advances, in *Quantitative Measurement of Physical Damage, ASTM STP 811*, Lankford, J., et al. eds. (American Society for Testing and Materials, Philadelphia), 464–484.
19. Purcell, A.H., and Weertman, J. (1974). Crack tip area in fatigued copper single crystals, *Metall. Trans.*, **5**, 1805–1809.
20. Hahn, G.T., and Rosenfield, A.R. (1968). Sources of fracture toughness: the relation between K...? and the ordinary tensile properties of metals, in *Applications Related Pkenomenn in Titanium Alloys, ASTM STP 432* (American Society for Testing and Materials, Philadelphia), 5–32.
21. Hahn, G.T., Hoagland, R.G., and Rosenfield, A.R. (1972). Local yielding attending fatigue crack growth, *Metall. Trans.*, **3**, 1189–1202.
22. Ranganathan, N. (1985). Contribution au développement d'une approche énergétique à la fissuration par fatigue (Thèse, Université de Poitiers, France).
23. Izumi, Y., and Fine, M.E. (1979). Role of plastic work in fatigue crack propagation in metals, *Eng. Fract. Mech.*, **11**, 791–804.
24. Iino, Y. (1976). Accumulated plastic zone around fatigue crack in type 304 stainless steel, *Metal. Sci.*, **10**, 159–164.
25. Tschegg, E., Faltin, C., and Stanzl, S. (1980). X-ray microbeam and recrystallization studies of plastic deformation around fatigue cracks, *J. Mater. Sci.*, **15**, 131–138.
26. Kwun, S.I., and Park, S.H. (1987). Plastic zone size measurement by critical grain growth method, *Scripta Metall.*, **21**(6), 797–800.

27. Lai, M.O., and Ferguson, W.G. (1980). Relationship between the shear lip size and the fracture toughness. *Mater. Sci. Eng.*, **45**, 183–188.
28. Hertzberg, R.W. (1989). *Deformation and Fracture Mechanics of Engineering Materials* (Wiley, Singapore), 608–610.
29. Davidson, D.L. (1984). Uses of electron channelling in studying material deformation, *Int. Metals Rev.*, **29**, 75–95.
30. Spencer, J.I., and Humphreys, C.J. (1980). A multiple scattering transport theory for electron channelling patterns, *Philos. Mag. A*, **42**, 433451.
31. Schulson, E.M. (1977). Electron channeling patterns in scanning electron microscopy. 1, *Mater. Sci.*, **12**, 1071–1087.
32. Joy, D.C., Newbury, D.E., and Davidson, D.L. (1982). Electron channelling patterns in the scanning electron microscope. 1, *Appl. Phys.*, **53**, R81–R122.
33. Stickler, R., Hughes, C.W., and Booker, G.R. (1971). Application of the SA-ECP method to deformation studies, in *Scanning Electron Microscopy/1971 (Part l), Proc. 4th Annual Scanning Electron Microscope Symposium* (ITT Research Institute, Chicago), 473–481.
34. Tekin, A., and Martin, J.W. (1989). High-resolution measurement of crack-tip plastic zone sizes by selected area channelling patterns, *Metallography*, **22**, 1–14.
35. Dung Do, T. (2013). Etude la zone plastique en tête d'une fissure de fatigue dans l'alliage 2024 T351 (Thèse, Université de Tours, France).

Chapter 6

Multiple Aspects of Polychloroprene's Fatigue Behavior

J. L. Poisson,[a,b] S. Meo,[b] F. Lacroix,[b] G. Berton,[b] and N. Ranganathan[b]

[a] *Tun Abdul Razak Research Center, Brickendonbury, Hertford, SG13 8NL, UK*
[b] *Laboratoire de Mécanique et Rhéologie (LMR) EA2640, Université François Rabelais de Tours (UFRT), CERMEL Center, 7 Avenue Marcel Dassault, 37200 Tours, France*
jpoisson@tarrc.co.uk, stephane.meo@univ-tours.fr

6.1 Introduction

Understanding phenomena induced by fatigue loading constitutes an important industrial issue. Therefore, two approaches were developed in the literature to investigate fatigue of materials, crack initiation and propagation. The second one is based on the work of Griffith [1], adapted to rubber-like materials by Rivlin and Thomas [2] and validated by Greensmith and Thomas [3], Greensmith [4], Lake and Lindley [5], Lindley [6], and Thomas [7]. More recent works were proposed by Mars [8, 9] and Mars and Fatemi [10, 11]. This chapter won't develop this approach. The other one consists in estimating fatigue lives of structures, without any prepared flaws

Materials Characterization: Modern Methods and Applications
Edited by Narayanaswami Ranganathan

ISBN 978-981-4613-06-4 (Hardcover), 978-981-4613-07-1 (eBook)
www.panstanford.com

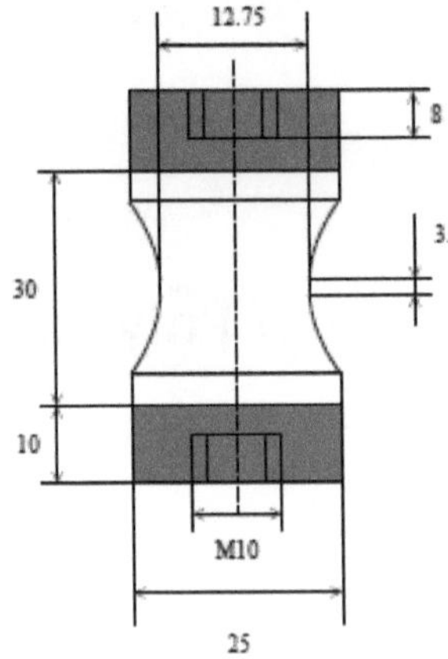

Figure 6.1 Dumbbell-type specimen.

or cracks. A parameter from the finite continuum mechanics, the fatigue criterion, is introduced as an estimator of fatigue lives. To validate such a method, an experimental investigation is required to confront the predictor with the experimental database.

Energy-based fatigue criteria have obtained an important expansion recently, with their scalar nature able to unify multiaxial results. The first one, the strain energy density (SED), was first investigated by Roberts and Benzies [12]. It was convenient because of its link with the tearing energy under certain conditions. However, Mars and Fatemi [13] pointed out the limits of this criterion in multiaxial cases. So they developed their own criterion, the cracking energy density (CED), which represents the SED available to the fatigue crack opening on the cracking plane. This parameter encountered real success in the literature [14–18]. More recently, from the work of Ellyin [19], who used the plastic energy density to determine the multiaxial fatigue behavior of metallic materials, Lacroix [20] and Lacroix et al. [21] have proposed the dissipated energy density (DED) as a criterion to describe the uniaxial fatigue behavior of polychloroprene rubber dumbbell specimens (Fig. 6.1).

Two methods were proposed to determine this parameter. The first approach, based on the experimental global results, defines the DED as the area contained on the stress–strain hysteresis loop resulting from the cyclic loading of rubber-like materials. And a local

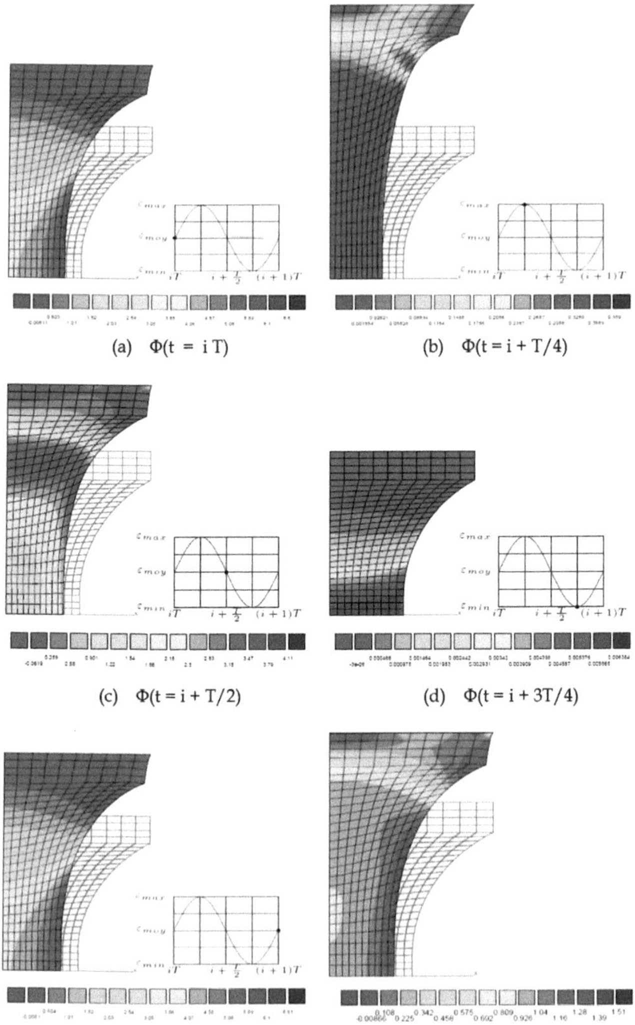

(a) Φ(t = i T)

(b) Φ(t = i + T/4)

(c) Φ(t = i + T/2)

(d) Φ(t = i + 3T/4)

(e) Φ(t = (i + 1)T)

(f) Average dissipation Φ(t = (i + 1)T)

Figure 6.2 Intrinsic dissipation evolution during a whole cycle.

calculation uses the finite element method to compute the intrinsic dissipation (Fig. 6.2). The latter parameter being unstable during a complete cycle, its average was then chosen as the fatigue criterion. Both criteria displayed encouraging results (Fig. 6.3).

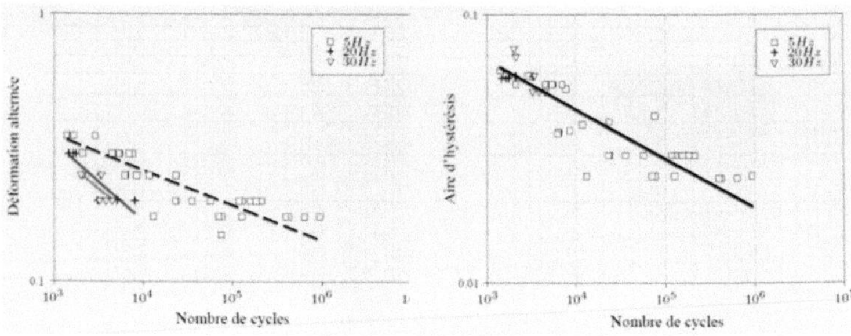

Figure 6.3 S–N curves. Tension loadings, frequency 5 Hz. From Refs. [20, 21].

An accurate estimation of a material's fatigue life is essential to the industrial parts' conception. Nevertheless, knowing the mechanisms leading to material rupture is indispensable to optimize materials' fatigue behavior. The determination of multiaxial fatigue behavior of structures with the fatigue criterion does not provide this information. In this purpose, many works [20, 22–27] were aimed to describe material fatigue damaging features with the fatigue failure surfaces observations realized with a scanning electron microscope. A complete bibliography of rubber fractography is available in the work of Poisson et al. [28].

The objective of this chapter is to present trends in the fatigue behavior description of polychloroprene rubber. The first part is intended to understand the material phenomena related to fatigue damaging, with scanning electron microscopy observations of fatigue failure surfaces. Then, a characterization of polychloroprene fatigue behavior is proposed in both uniaxial and multiaxial conditions. This description is realized in the crack initiation frame. The DED, proposed by Lacroix [20] and Lacroix et al. [21], is tested first in uniaxial loading and confronted with other criteria in multiaxial solicitations.

6.2 Polychloroprene Fatigue-Damaging Features

The observation of tension fatigue failure surfaces is realized to determine the impact of load ratio ($R = F_{\min}/F_{\max}$) on polychloroprene

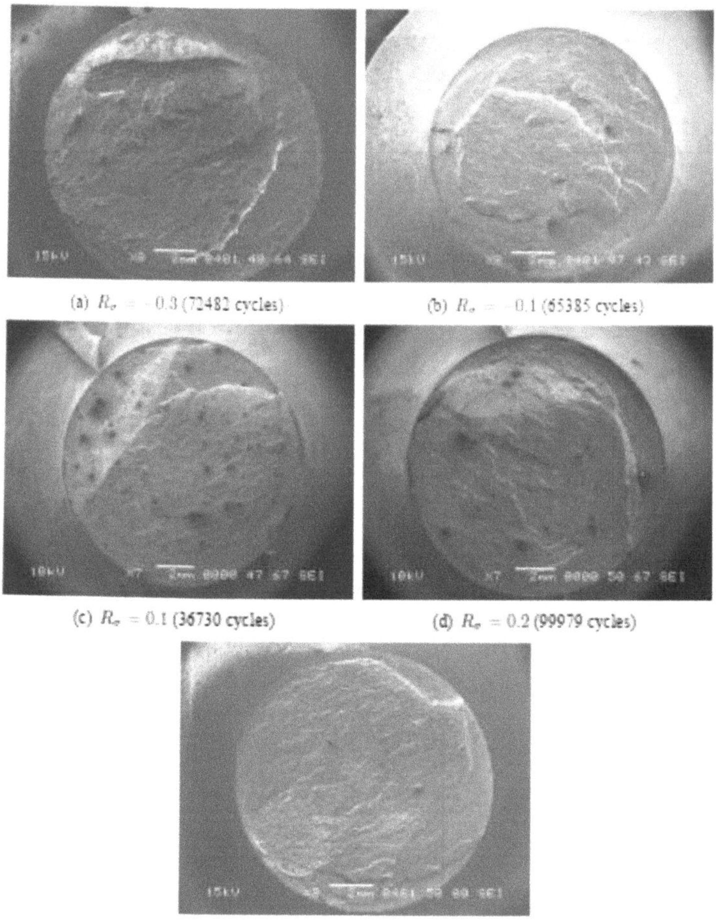

(a) $R_\sigma = -0.3$ (72482 cycles)
(b) $R_\sigma = -0.1$ (65385 cycles)
(c) $R_\sigma = 0.1$ (36730 cycles)
(d) $R_\sigma = 0.2$ (99979 cycles)
(e) $R_\sigma = 0.4$ (491075 cycles)

Figure 6.4 Scanning electron microscope observation, at a macroscopic scale, of specimen fatigue failure surfaces.

fatigue damaging, the maximum load being fixed to 150 N. As a first step, analyses are carried out at a macroscopic scale to detect the load ratio effect on the global failure surface morphology (Fig. 6.4). This work is managed with a scanning electron microscope JEOL JSM 5900 LV. For all the specimens subjected to positive load ratios, rupture surfaces can be divided into two different regions, a rough

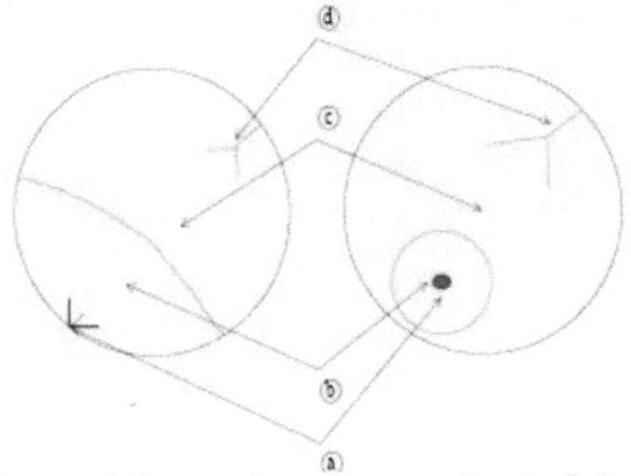

(a) Fatigue failure surface scheme and main fatigue zones

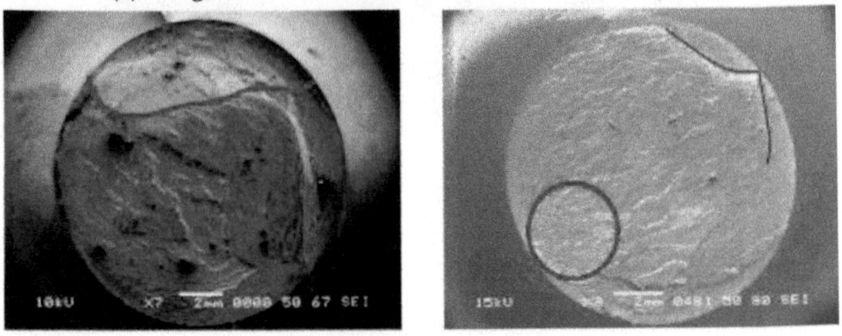

(b) 2 examples of fatigue failure surfaces

Figure 6.5 Fatigue failure surfaces of specimen subjected to positive loading ratio tension loading: (a) crack initiation zone, (b) stable crack growth zone, (c) end-of-life zone, and (d) tearing zone.

zone representing the stable crack growth area and a lighter domain corresponding to the end-of-life zone (Fig. 6.5).

In this work, only the crack initiation zone related to fatigue solicitations with a positive load ratio will be considered. Readers are invited to read the thesis of Poisson [29] to get a description of the tearing zone.

When attention is focused on the crack initiation zone relative to four tension fatigue failure surfaces, three of them present a crack initiated on the molding joint (Fig. 6.6a–c) and around an internal flaw for the other one (Fig. 6.6d). Figure 6.7 displays the equivalent of Fig. 6.6, with the chemical contrast highlighting the presence of black spots. A zoom of them, displayed in Fig. 6.8, reveals tongues, similar to those observed in the literature (by Legorju-Jago with the natural rubber [22], Lacroix with the polychloroprene rubber [20] and Gauchet with the hydrogenated nitrile butadiene

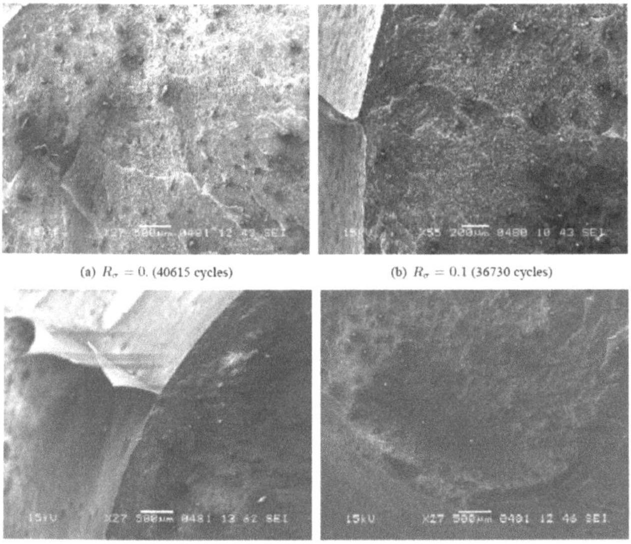

(a) $R_\sigma = 0$. (40615 cycles)

(b) $R_\sigma = 0.1$ (36730 cycles)

(c) $R_\sigma = 0.2$ (99979 cycles)

(d) $R_\sigma = 0.4$ (491075 cycles)

Figure 6.6 Fatigue failure surfaces of specimens subjected to positive loading ratio tension solicitations.

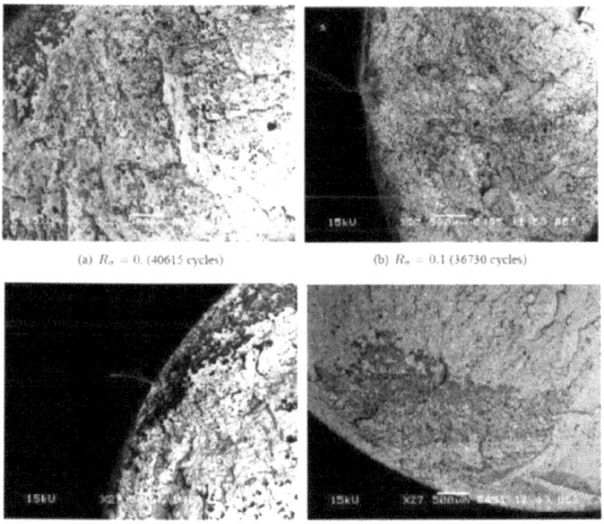

(a) $R_\sigma = 0$. (40615 cycles)

(b) $R_\sigma = 0.1$ (36730 cycles)

(c) $R_\sigma = 0.2$ (99979 cycles)

(d) $R_\sigma = 0.4$ (491075 cycles)

Figure 6.7 Fatigue failure surfaces of specimens subjected to positive loading ratio tension solicitations.

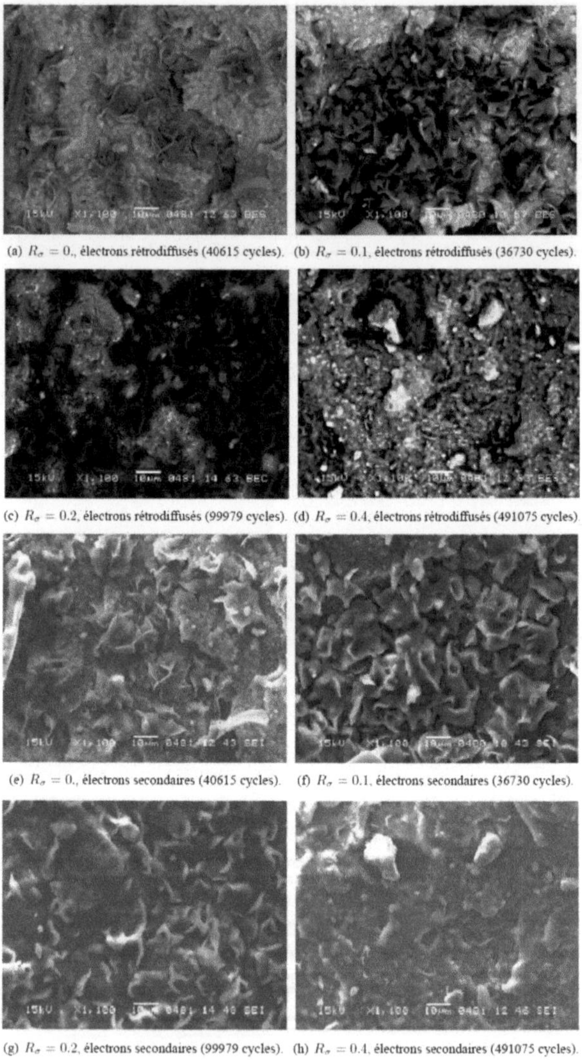

(a) $R_\sigma = 0.$, électrons rétrodiffusés (40615 cycles). (b) $R_\sigma = 0.1$, électrons rétrodiffusés (36730 cycles).

(c) $R_\sigma = 0.2$, électrons rétrodiffusés (99979 cycles). (d) $R_\sigma = 0.4$, électrons rétrodiffusés (491075 cycles).

(e) $R_\sigma = 0.$, électrons secondaires (40615 cycles). (f) $R_\sigma = 0.1$, électrons secondaires (36730 cycles).

(g) $R_\sigma = 0.2$, électrons secondaires (99979 cycles). (h) $R_\sigma = 0.4$, électrons secondaires (491075 cycles).

Figure 6.8 SEM observations of the crack initiation zone (×1100), revealing the tongues' morphologies.

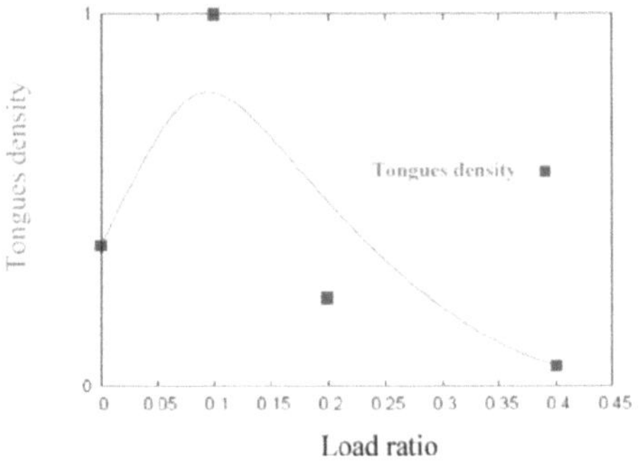

Figure 6.9 Impact of the loading ratio on the density of tongues on the crack initiation zone.

rubber [23]). Besides, when the pictures of Fig. 6.8 are intensively observed, a difference can be noted on the tongues' morphology, getting smoother by increasing the load ratio.

Legorju-Jago [30] concluded that these tongues represented the thermo-oxidation of a rubber matrix under cyclic loading. An analysis of Fig. 6.7 is carried out to qualify the black spot density, which means the tongues' density on the stable crack growth zone. Figure 6.9 exposes the impact of loading ratio on the tongues' density. It reveals a change in the tongues' density from a load ratio of about 0.2 and suggests a change in material fatigue damaging. A correlation with the fatigue lives will be proposed later in this chapter. Further analyses and observations are yet needed to clearly state the link between tongues and fatigue damaging.

In this part, a macroscopic description of specimen failure surfaces subjected to tension fatigue loading has been proposed, with two main zones, the stable crack growth and the end-of-life zones. A focus has been made on the first region. A retrodiffused analysis, realized with a scanning electron microscope, has pointed out black spots, which turned out to be tongues. These features seemed to get smoother morphologies by increasing the load ratio. Qualitative determination of the tongues' density on the crack initiation zone has revealed a possible change in fatigue damaging at a load ratio of about 0.2. First trends have been presented in this part to describe the polychloroprene tension fatigue damaging. This work will be useful to get deeper in the description of this material fatigue damaging. Nevertheless, this part does not provide any clues to estimate the fatigue life of rubber products or to set a link between fatigue lives and the tongues' density. That will be the aim of the next part.

6.3 Polychloroprene Fatigue Behavior

In this section, the fatigue crack initiation behavior of polychloroprene rubber is described. The first part is dedicated to uniaxial tension fatigue characterization of the material. Then, fatigue criteria are tested in multiaxial conditions in the second part.

6.3.1 *Uniaxial Fatigue*

An experimental tension fatigue campaign has been realized by Poisson et al. [29, 31] to observe the impact of the loading ratio on the tension fatigue behavior of the CR rubber. These experiments were conducted with an INSTRON 8802 fatigue-testing device, load-driven at a frequency of 5 Hz, at room temperature, and the tests conditions are summarized in Table 6.1. Besides, this investigation is an opportunity to test the DED as a fatigue estimator. Thus, only simple fatigue criteria are explored in this part: the stress maximum (Eq. 6.1), the stress amplitude (Eq. 6.2), the SED (Eq. 6.3), and the DED (Eq. 6.4).

$$\pi_{\max} = \max_{\text{cycle}} \pi(t) \text{ with } \pi(t) = \frac{F(t)}{S_0} \tag{6.1}$$

Table 6.1 Tension fatigue experimental conditions

$F_{\max}$	R	$F_{\max}$	R	$F_{\max}$	R	$F_{\max}$	R	$F_{\max}$	R
					0.5		0.4		
	0.5		0.5						0.3
					0.4		0.3	160 N,	
300 N									
	0.4		0.4		0.3		0.2		0.2
			0		0.2		0.1		0.1
		200 N		175 N		150 N			
					0.1		0	125 N,	0
	0.5		−0.1						
					0		−0.1		−0.1
250 N									
					−0.1				
	0.4		−0.3				−0.3	100 N	−0.3
					−0.3				

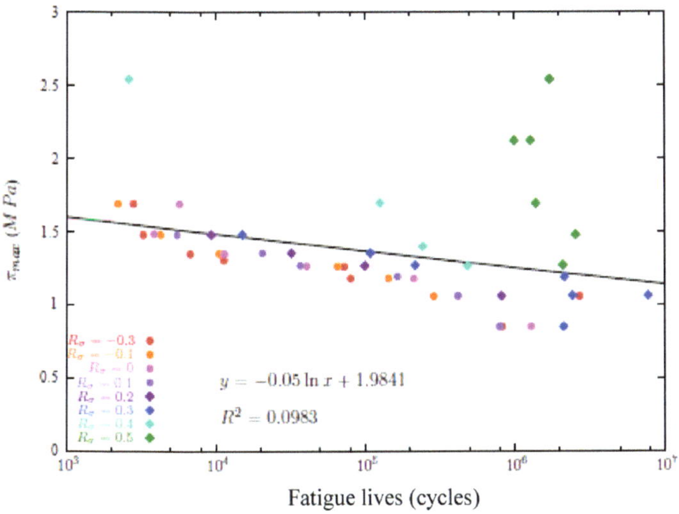

Figure 6.10 Uniaxial tension fatigue results of polychloroprene with maximum stress.

$$\pi_a = \frac{1}{2}(\pi_{max} - \pi_{min}) \text{ with } \pi_{min} = \min_{\text{cycle}} \pi(t) \tag{6.2}$$

$$\phi = \min_{\text{cycle}} W(t) \text{ with } W(t) = \int_t \pi(t)d\varepsilon \tag{6.3}$$

$$\Phi = \int_{\text{cycle}} \pi(t)d\varepsilon \tag{6.4}$$

F, S_0, and ε are, respectively, the imposed load, the initial central section of the specimen, and the axial strain.

All the criteria were determined right before rupture. Moreover, the end-of-life criterion is considered, in this work, as the total rupture of the specimen.

Figures 6.10 to 6.13 display the tension fatigue results of polychloroprene with, respectively, the maximum stress, the stress amplitude, the SED, and the DED. Fatigue results are displayed as points, the straight line represents the global correlation, and the R^2 coefficient reveals its accuracy. At first sight, for stress maximum and SED criteria, it seems like the $R = 0.4$ and $R = 0.5$ fatigue results are far away from the other results. These weird results caused a bad correlation for the concerned criteria. Meanwhile,

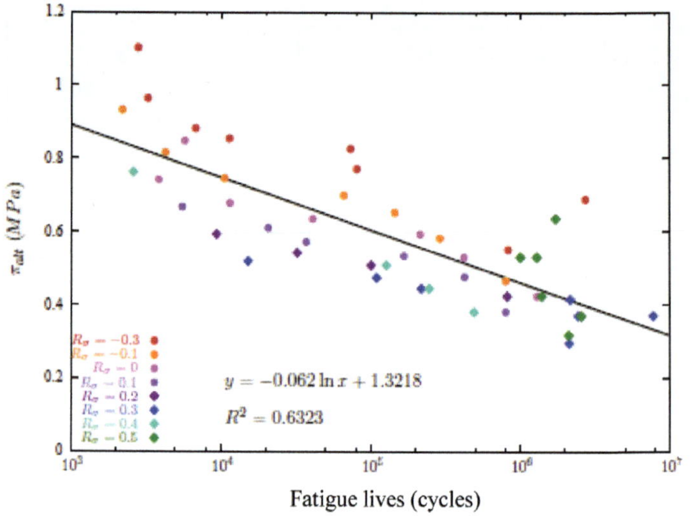

Figure 6.11 Uniaxial tension fatigue results of polychloroprene with stress amplitude.

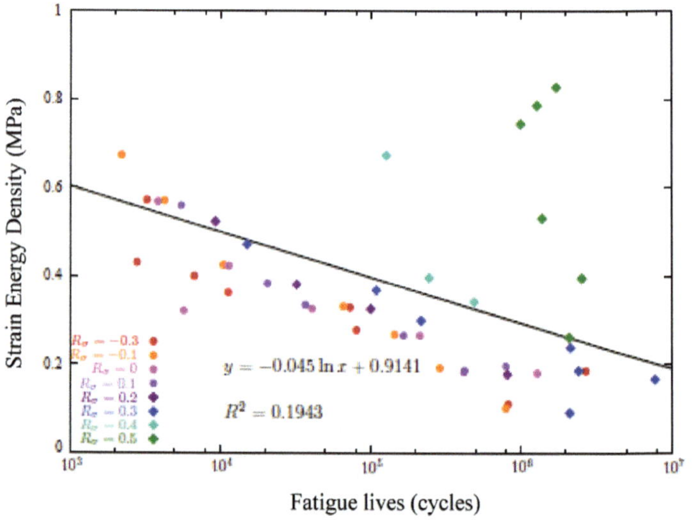

Figure 6.12 Uniaxial tension fatigue results of polychloroprene with strain energy density.

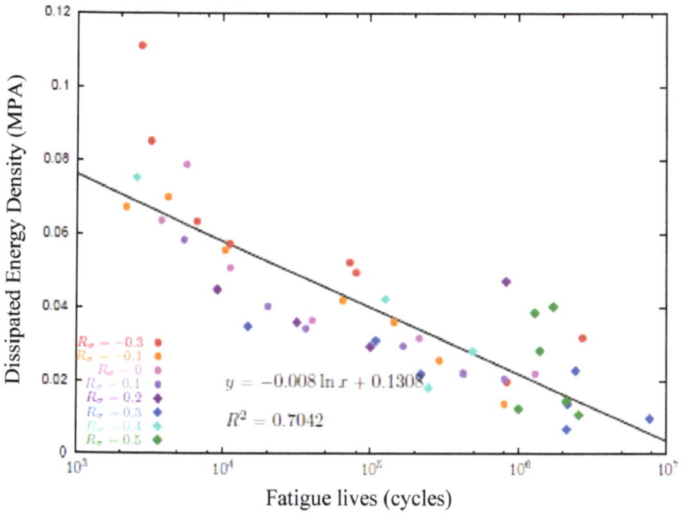

Figure 6.13 Uniaxial tension fatigue results of polychloroprene with dissipated energy density.

stress amplitude and the DED displayed encouraging results. It appears that load ratio has got an important impact on the fatigue results. To quantify the effect of load ratio on the fatigue behavior of natural rubber, many researchers have built a Haigh diagram [11, 32–35], a graph where fatigue results are gathered as points, whose abscissa and ordinate are, respectively, mean stress and stress amplitude. Every fatigue life is indexed, so iso-fatigue life curves can be plotted. The objective of such representation is to describe the fatigue behavior of rubber-like materials with two parameters. These previous works highlighted the reinforcing nature of natural rubber fatigue behavior under a positive load ratio. They explained this phenomenon with the crystallization under a tension state of the material. The polychloroprene Haigh diagram is displayed in Fig. 6.14.

Figure 6.14 confirms this influence of load ratio on tension fatigue behavior of polychloroprene rubber. This results displays a reinforcing fatigue behavior of this material from $R_\sigma = 0.2$. This change in fatigue results is similar to the one observed with natural rubber. This can be explained by the appearance of crystallization

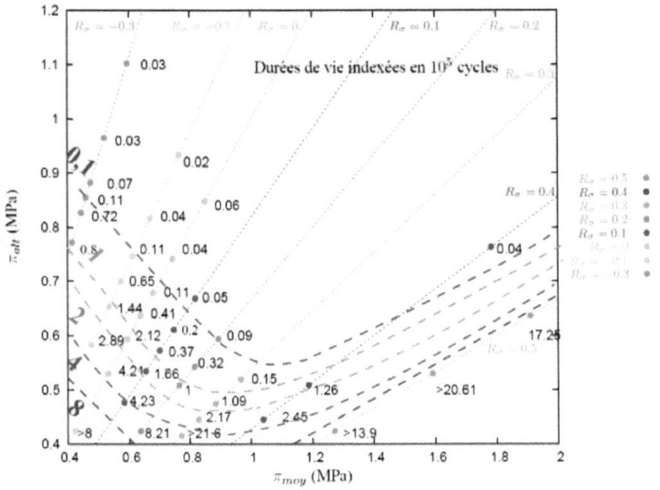

Figure 6.14 Tension fatigue Haigh diagram for polychloroprene rubber.

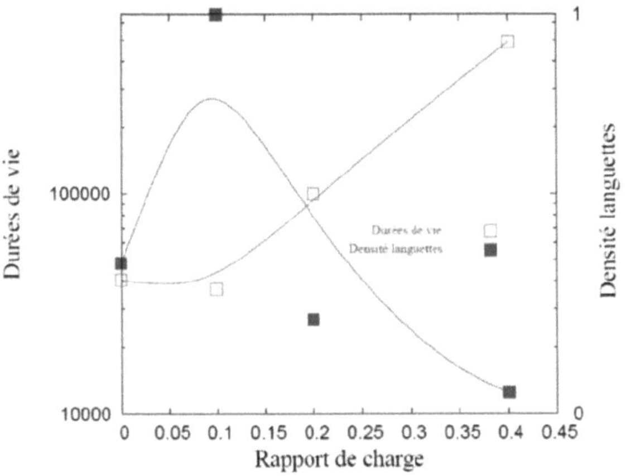

Figure 6.15 Comparison between fatigue lives and the tongues' density: impact of load ratio.

under a tension state. Besides, when the fatigue life results are compared with the tongues' density plotted in Fig. 6.9 (Fig. 6.15), a change is observed for both results at the same time. This result gives the evidence that tongues are closely related to material fatigue damaging.

Experimental uniaxial fatigue investigations on polychloroprene have tested many simple criteria. It appears that the DED exposed the most interesting results, by taking into account the phenomena related to the load ratio. Besides, the Haigh diagram displayed results quite similar to those observed with natural rubber, raising the possibility of crystallization under a tension state. Indeed, Toki [36] highlighted a link between hysteresis energy and crystallization under tension. This explains the interesting results of the DED as a tension fatigue criterion. Let's see now if this change in fatigue damaging is also appearing in multiaxial solicitations.

6.3.2 *Muliaxial Fatigue*

To complete the observations realized in tension loading, combined tension–torsion solicitations are carried out with an electro-force fatigue testing device BOSE3300. These experiments are displacement–rotation driven. The experimental conditions are listed in Table 6.2.

This section is aimed to confront the DED with different principal multiaxial fatigue criteria available in the literature. To calculate these criteria, it is compulsory to determine the stress–strain state in our structure. Three approaches are chosen in that purpose:

Table 6.2 Multiaxial fatigue experimental conditions

δ (°)	R_ε	U_{max} (mm)	U_{min} (mm)	θ_{max} (°)	θ_{min} (°)
0° , 90°, 180° tension	0.2	20	4	40	8
		17	3.4	34	6.8
		15	3	30	6
	0.1	20	2	40	4
		17	1.7	34	3.4
		15	1.5	30	3
	0	20	0	40	0
		17	0	34	0
		15	0	30	0
	−0,2	20	−4	40	−8
		17	−3.4	34	−6.8
		15	−3	30	−6

- An analytic method, developed by Poisson et al. [31, 37] and by Poisson [29] that consists of calculating the stress–strain state at one point with a large strain Kelvin–Voigt model (local).
- A finite element method, implemented in ANSYS software, that estimates mechanical parameters in the whole structure [29]. The fatigue criteria are then calculated at one node of the structure corresponding to midheight (local).
- A global method that uses directly the raw experimental data.

The fatigue criteria, which are computed from a stabilized fatigue cycle and investigated in this section, are:

- The first principal stress (PK1), from the first Piola–Kirchhoff stress tensor. It was shown by Saintier [34] and Saintier et al. [38] that the crack opening is driven by the first principal stress. It is also the most intuitive and simple criterion.
- The maximum of the SED, defined globally (Eq. 6.5) and locally (Eq. 6.6):

$$\phi_{\text{global}} = \max_{\text{cycle}} \left(W^{\text{g}}(t)\right) \text{ and } W^{\text{g}}(t) = \frac{1}{V_{\text{h}}} \int_t F\,dU + C\,d\theta \tag{6.5}$$

$$\phi_{\text{local}} = \max_{\text{cycle}} \left(W^{\text{l}}(t)\right) \text{ and } W^{\text{l}}(t) = \int_t \overline{\overline{\pi}} : d\overline{\overline{F}} \tag{6.6}$$

- The DED, defined globally (Eq. 6.7) and locally (Eq. 6.8):

$$\Phi^{\text{g}}(t) = \frac{1}{V} \int_{\text{cycle}} F\,dU + C\,d\theta \tag{6.7}$$

$$\Phi^{l}(t) = \int_{\text{cycle}} \overline{\overline{\pi}} : d\overline{\overline{F}} \tag{6.8}$$

  where $\overline{\overline{F}}$ corresponds to the gradient tensor.
- The Eshelby criterion (Σ^*), based on the configurational mechanics, is developed in the literature by Verron et al. [39], Andriyana and Verron [40], Verron and Andriyana [41], Andriyana et al. [42], and Previati and Kaliske [43]. This parameter is calculated from Eq. 6.9:

$$\Sigma^* = \left|\min\left((\Sigma_i)_{i=1,2,3}, 0\right)\right| \tag{6.9}$$

where Σ_i are corresponding to the eigenvalues of the Eshelby stress tensor $\overline{\overline{\Sigma}}$, defined by Eq. 6.10 from the SED W, the right Cauchy–Green strain tensor $\overline{\overline{C}}$, and the second Piola–Kirchhoff stress tensor $\overline{\overline{S}}$:

$$\overline{\overline{\Sigma}} = W\overline{\overline{I}} - \overline{\overline{C}}.\overline{\overline{S}} \tag{6.10}$$

To take into account the loading history in the formulation of their criterion, Andriyana and Verron [40] decided to differentiate it (Eq. 6.11).

$$d\overline{\overline{\Sigma}} = \frac{d\overline{\overline{\Sigma}}}{d\overline{\overline{C}}} : d\overline{\overline{C}} = \sum_{i=1}^{3} d\Sigma_i \overline{\mathbf{V}}_i \otimes \overline{\mathbf{V}}_i \tag{6.11}$$

where $d\Sigma_i$ is the i-th eigenvalue of the stress tensor $d\overline{\overline{\Sigma}}$ and $\overline{\mathbf{V}}_i$ its corresponding eigenvector. Andriyana and Verron [40] noticed that only negative values of their criterion caused fatigue damages because of crack-closing issues. Thus, they introduced a damaging part of their criterion (Eq. 6.12).

$$d\overline{\overline{\Sigma}}^d = \sum_{i=1}^{3} d\Sigma_i^d \overline{\mathbf{V}}_i \otimes \overline{\mathbf{V}}_i \tag{6.12}$$

with

$$d\Sigma_i^d = \begin{cases} d\Sigma_i \; if \;\; d\Sigma_i < 0 \text{ and } \overline{\mathbf{V}}_i.\overline{\overline{\Sigma}}.\overline{\mathbf{V}}_i < 0 \\ 0 \text{ otherwise} \end{cases} \tag{6.13}$$

Finally, the modified fatigue criterion is expressed in Eq. 6.14.

$$\Sigma^* = \left|\min\left((\Sigma_i)_{i=1,2,3}, 0\right)\right| \tag{6.14}$$

- The last criterion explored is the CED, whose formulation is given in Eq. 6.15.

$$\text{CED} = \int_{\text{cycle}} \frac{\overline{\mathbf{R}}^T.\overline{\overline{C}}.\overline{\overline{S}}.d\overline{\overline{E}}.\overline{\mathbf{R}}}{\overline{\mathbf{R}}^T.\overline{\overline{C}}.\overline{\mathbf{R}}} \tag{6.15}$$

where $\overline{\mathbf{R}}$ represents the vector normal to the cracking plane.

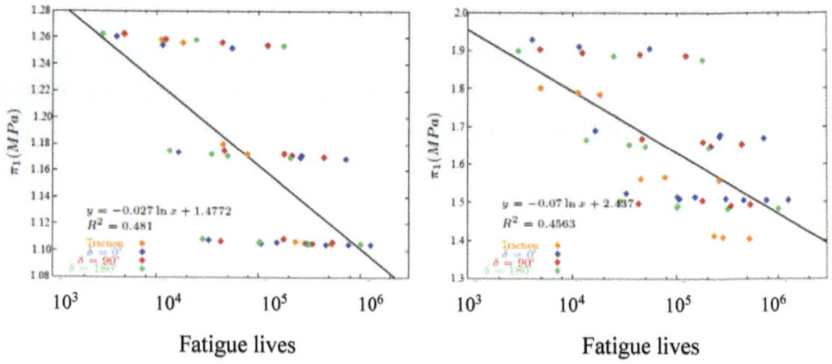

(a) Analytic calculation (b) Finite Element calculation

Figure 6.16 Multiaxial fatigue results with the PK1 criterion.

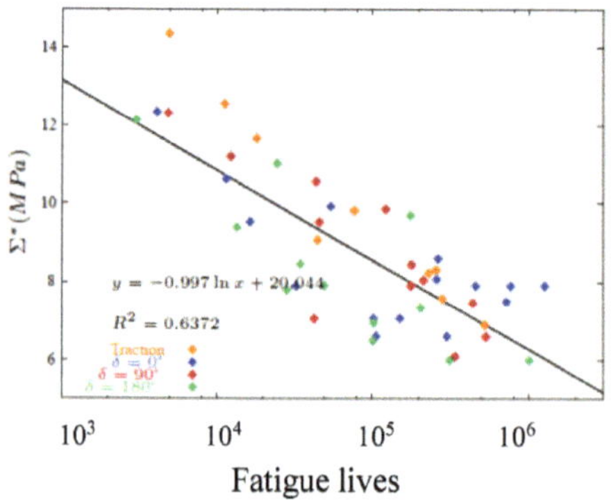

Figure 6.17 Multiaxial fatigue results with the Σ* criterion (analytic calculation).

Figures 6.16 to 6.20 give the representations of the criteria analyzed in this section. Figures 6.16a and 6.16b point out the fact that the local PK1 criterion is not adapted to describe multiaxial fatigue behavior of polychloroprene. It can be explained by its formulation that does not take into account the loading history. The Eshelby criterion (Fig. 6.17) seems to fit better, due probably to its partial time integration, so as the SED criterion (Fig. 6.18).

Concerning both DED and CED criteria, they exposed the best results, because of their total time integration of the cycle, with best correlations for the DED criterion. Besides, with this latter

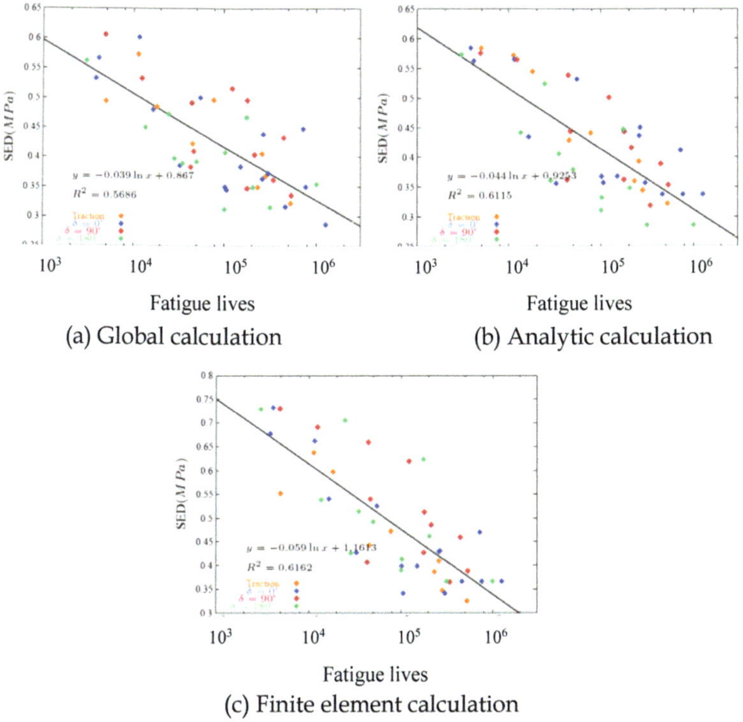

Figure 6.18 Multiaxial fatigue results with the SED criterion.

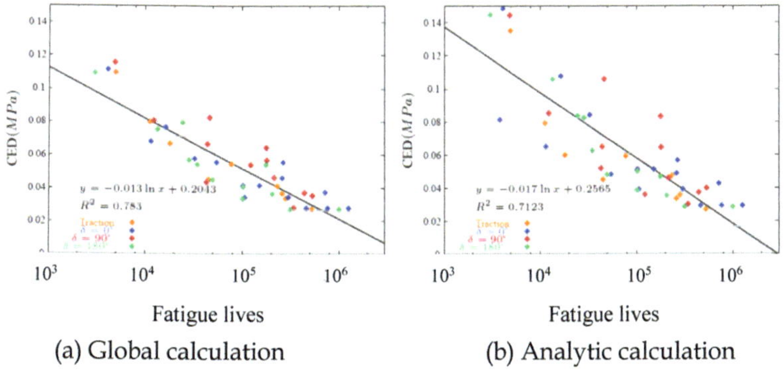

Figure 6.19 Multiaxial fatigue results with the CED criterion.

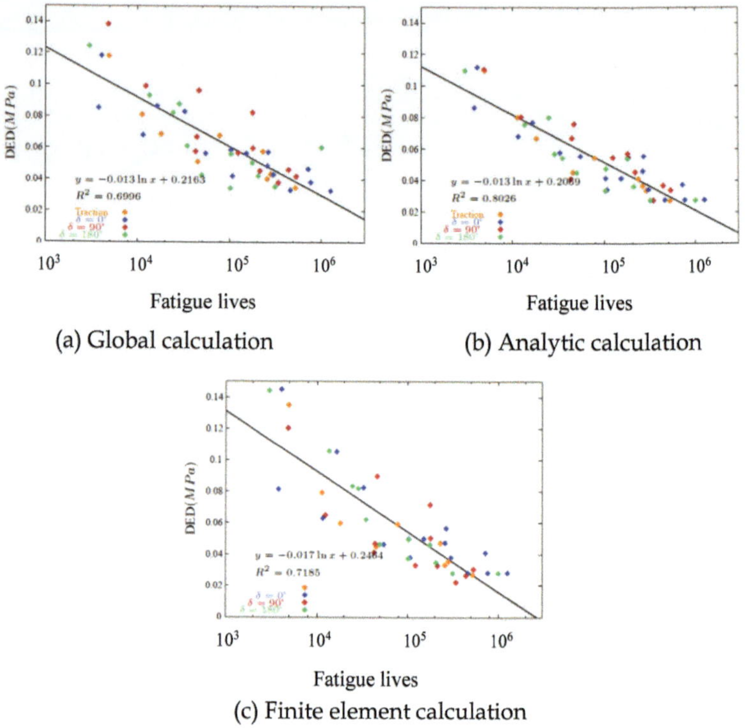

Figure 6.20 Multiaxial fatigue results with the DED criterion.

parameter, it is to be noticed that global and local calculations have led to close results. This statement validates both approaches. It is then natural to choose the faster method to determine polychloroprene fatigue behavior in the future (analytic, global). Nevertheless, the finite element method would be still required in order to work with complex structures.

Since the DED criterion is the most adapted parameter to unify multiaxial fatigue behavior of polychloroprene, it is then consistent to choose this value as the amplitude term of a multiaxial Haigh diagram, completed with the mean hydrostatic pressure (Eq. 6.16) to represent the mean part.

$$p_{\text{mean}} = \frac{1}{2}\left(\max_{\text{cycle}}\left(p(t)\right) + \min_{\text{cycle}}\left(p(t)\right)\right) \tag{6.16}$$

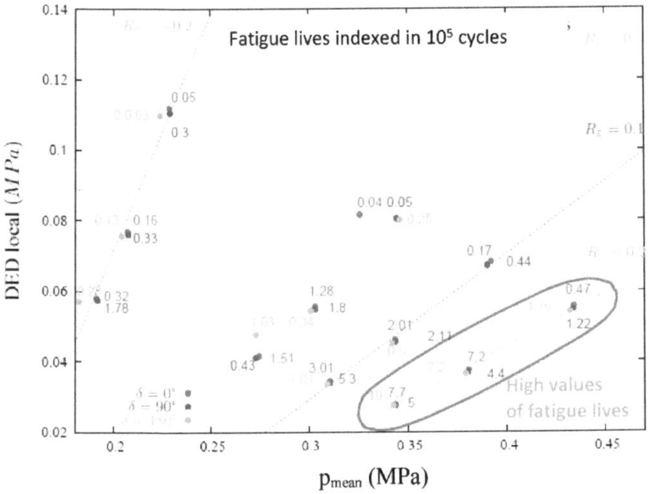

Figure 6.21 Multiaxial Haigh diagram.

Figure 6.21 displays the multiaxial Haigh diagram of polychloroprene rubber. This figure, even if a completion is needed, reveals one more time a reinforcing phenomenon from a loading ratio of about 0.2. This result implies the existence of crystallization under a multiaxial state, which has not been ever shown in the literature. Nevertheless, additional results are needed to confirm this trend.

6.4 Conclusion

This investigation, aimed at studying the fatigue behavior of polychloroprene rubber, was realized with the crack initiation approach, implying the definition of a criterion, and the use of "healthy" specimens. Scanning electron microscopy observations have led to a description of macroscale fatigue failure surfaces divided into two zones:

- Stable crack growth, a rough region
- End of life, a lighter domain

The first region, being focused, microscale analyses revealed fatigue-damaging features, tongues, whose density on the whole crack initiation zone is related to the fatigue damaging of polychloroprene rubber.

The description of its uniaxial fatigue behavior pointed out a reinforcing phenomenon from a load ratio of about 0.2, similar to that observed with natural rubber and explained by crystallization under a tension state. This kink is consistent with the results observed with a microscope, confirming the link between fatigue damaging and tongues. Moreover, in spite of the SED and maximum stress, the DED managed to describe well the tension fatigue behavior of polychloroprene rubber, taking into account the phenomenon related to load ratio.

This criterion was then confronted, in multiaxial solicitations, to several parameters available in the literature. With the CED, the DED obtained the most interesting results, explained by the integration of the whole cycle in their formulations. This parameter was therefore used to describe multiaxial fatigue behavior of polychloroprene rubber, with a Haigh diagram. This diagram suggested a difference in fatigue damages for a 0.2 R_d ratio, which means the possibility of crystallization of this material in multiaxial loadings. This result is consistent with the tension fatigue Haigh diagram displayed by Poisson et al. [29] for polychloroprene. Besides, short fatigue lives are observed for a negative R_d ratio, which differed from the hypothesis of Brunac et al. [44] when they investigated the Haigh diagram. Nevertheless, further investigations are required to confirm this tendency.

References

1. Griffith, A.A. (1921). The phenomena of rupture flow in solids, *Philos. Trans. R. Soc. A*, **221**, 1028–1039.
2. Rivlin, R.S., and Thomas, A.G. (1953). Rupture of rubber I. Characteristic energy for tearing, *J. Polym. Sci.*, **10**, 291–318.
3. Greensmith, H.W., and Thomas, A.G. (1955). Rupture of rubber III. Determination of tear properties, *J. Polym. Sci.*, **18**, 189–200.
4. Greensmith, H.W. (1960). Rupture of rubber VII. Effect of rate extension in tensile tests, *J. Appl. Polym. Sci.*, **3**, 175–182.

5. Lake, G.J., and Lindley, P.B. (1965). Cut growth and fatigue of rubbers II. Experiments on a noncrystallizing rubber, *Rubber Chem. Technol.*, **38**, 292–300.
6. Lindley, P.B. (1972). Energy for crack growth in model rubber components, *J. Strain Anal. Eng.*, **7**, 132–140.
7. Thomas, A.G. (1960). Rupture of rubber VI. Further experiments on tear criterion, *J. Appl. Polym. Sci.*, **3**, 168–174.
8. Mars, W.V. (2001). Multiaxial fatigue of rubber (PhD thesis, University of Toledo, USA).
9. Mars, W.V. (2003). A phenomenological model for the effect of R: ratio on fatigue of strain crystallizing rubber, *Rubber Chem. Technol.*, **76**, 1240–1258.
10. Mars, W.V., and Fatemi, A. (2007). The correlation of fatigue crack growth rates in rubber subjected to multiaxial loading using continuum mechanical parameters, *Rubber Chem. Technol.*, **80**, 169–182.
11. Mars, W.V., and Fatemi, A. (2009). Computed dependence of rubber's fatigue behavior on strain crystallization, *Rubber Chem. Technol.*, **82**, 51–61.
12. Robers, B.J., and Benzies, J.B. (1977). The relationship between uniaxial and equibiaxial fatigue in gum and carbon black filled vulcanizates, *Proc. Rubbercon, 77*, **2**, 2–13.
13. Mars, W.V., and Fatemi, A. (2002). A littérature survey on analysis approaches for rubber, *Int. J. Fatigue*, **24**, 949–969.
14. Zine, A., Benseddiq, N., and Nait Abdelaziz, M. (2006). Prediction of rubber fatigue life under multiaxial loading, *Fatigue Fract. Eng. Mater. Struct.*, **29**, 267–278.
15. Zine, A., Benseddiq, N., and Nait Abdelaziz, M. (2011). Rubber fatigue life under multiaxial loading: numerical and experimental investigations, *Int. J. Fatigue*, **33**, 1360–1368.
16. Kim, T.W., Jeong, H.Y., Choe, J.H., and Kim, Y.H. (2005). Prediction of the fatigue of tires using CED and VCCT, *Key Eng. Mater.*, **297–300**, 102–107.
17. Harbour, R.J., Fatemi, A., and Mars, W.V. (2007). Fatigue crack growth of filled rubber under constant and variable amplitude loading conditions, *Fatigue Fract. Eng. Mater. Struct.*, **30**, 640–652.
18. Harbour, R.J., Fatemi, A. and Mars, W.V. (2008). Fatigue life and predictions for NR and SBR under variable amplitude and multiaxial loading conditions, *Int. J. Fatigue*, **30**, 1231–1247.
19. Ellyin, F. (1974). A criterion for fatigue under multiaxial state of stress, *Mech. Res. Commun.*, **1**, 219–224.

20. Lacroix, F. (2004). Endommagement en fatigue d'élastomères synthétiques (Thèse, Université François Rabelais de Tours, France).
21. Lacroix, F., Méo, S., Berton, G., Tougui, A. and Ranganathan, N. (2005). A local criterion for fatigue crack initiation on a vulcanized filled elastomer, in *Constitutive Models for Rubber*, Austrell, P.E., and Kari, L., eds. (A.A. Balkema), 77–81.
22. Legorju, K. (1998). Influence de la cristallisation sous contrainte sur l'endurance et la propagation de fissure de fatigue dans les élastomères (Thèse, Ecole Centrale Paris, France).
23. Gauchet, S. (2007). Etude de l'influence du type de noir de carbone sur la tenue en fatigue du caoutchouc HNBR (Thèse, Université François Rabelais de Tours, France).
24. Gauchet, S., Legorju-Jago, K., and Ranganathan, N. (2007). Influence du noir de carbone sur les mécanismes de rupture par fatigue de caoutchouc HNBR (Congrès Français de Mécanique).
25. Le Cam, J.B., Huneau, B., Verron, E. (2008). Description of fatigue damage in carbon black filled natural rubber, *Fatigue Fract. Eng. Mater. Struct.*, **31**, 1031–1038.
26. Bhowmick, A.K., Basu, S., and De, S.K. (1979). Scanning electron microscopy studies of nitrile rubber fractured surfaces, *Rubber Chem. Technol.*, **53**, 321–326.
27. Bhowmick, A.K., Nando, G.B., Basu, S., and De, S.K. (1979). Scanning electron microscopy studies of fractured natural rubber surfaces, *Rubber Chem. Technol.*, **53**, 327–334.
28. Poisson, J.-L., Lacroix, F., Méo, S., Berton, G., Ranganathan, N. (2012). Rubber fatigue: the intrinsic intricacies, in *Fatigue of Materials II: Advances and Emergences in Understanding*, Srivatsan, T.S., Imam, M.A., and Srinivasan, R., eds. (John Wiley and Sons, Hoboken, NJ).
29. Poisson, J.L. (2012). Détermination d'un critère de fatigue appliqué à un élastomère synthétique (Thèse, Université François Rabelais de Tours, France).
30. Legorju-jago, K., and Bathias, C. (2002). Fatigue initiation and propagation in natural and synthetic rubbers, *Int. J. Fatigue*, **24**, 85–92.
31. Poisson, J.L., Lacroix, F., Méo, S., Berton, G., and Ranganathan, N. (2011). Biaxial fatigue behavior of a polychloroprene rubber, *Int. J. Fatigue*, **33**, 1151–1157.
32. André, N. (1998). Critère local d'amorçage de fissure en fatigue dans un élastomère de type NR (Thèse, Ecole Nationale des Mines de Paris, France).

33. André, N., Cailletaud, G., and Piques, R. (1998). Haigh diagram for fatigue crack initiation prediction of natural rubber components, *Kautsch. Gummi Kunstst.*, **52**, 120–123.

34. Saintier, N. (1998). Fatigue multiaxiale dans un élastomère de type NR chargé: mécanismes d'endommagement et critère local d'amorçage de fissure (Thèse, Ecole Nationale des Mines de Paris, France).

35. Saintier, N., Cailletaud, G., and Piques, R. (2006). Multiaxial fatigue life prediction for a natural rubber, *Int. J. Fatigue*, **28**, 530–539.

36. Toki, S., Fujimaki, T., and Okuyama, M. (2000). Strain induced crystallization of natural rubbers as detected real time by wide angle x-ray diffraction technique, *Polymer*, **41**, 5423–5429.

37. Poisson, J.L., Méo, S., Lacroix, F., Berton, G., and Ranganathan, N. (2012). Multiaxial fatigue criteria applied to a polychloroprene rubber, *Rubber Chem. Technol.*, **85**, 80–91.

38. Saintier, N., Cailletaud, G., and Piques, R. (2006). Crack initiation and propagation under multiaxial fatigue in a natural rubber, *Int. J. Fatigue*, **28**, 61–72.

39. Verron, E., Le Cam, J.B., and Gornet, L. (2006). A multiaxial criterion for crack nucleation in rubber, *Mech. Res. Commun.*, **33**, 493–498.

40. Andriyana, A., and Verron, E. (2007). Prediction of fatigue life improvement in natural rubber using configurational stress, *Int. J. Solids Struct.*, **44**, 2079–2092.

41. Verron, E., and Andriyana, A. (2008). Definition of a new predictor for multiaxial fatigue crack nucleation in rubber, *J. Mech. Phys. Solids*, **56**, 417–498.

42. Andriyana, A., Saintier, N., and Verron, E. (2010). Configurational mechanics and critical plane approach: concept and application to fatigue failure analysis of rubberlike materials, *Int. J. Fatigue*, **32**, 1627–1638.

43. Previati, G., and Kaliske, M. (2012). Crack propagation in pneumatic tires: continuum mechanics and fracture mechanics approaches, *Int. J. Fatigue*, **37**, 69–78.

44. Brunac, J.B., Gerardin, O., and Leblond, J.B. (2009). On the heuristic extension of Haigh's diagram for the fatigue of elastomers to arbitrary loadings, *Int. J. Fatigue*, **31**, 859–867.

Chapter 7

Fatigue Crack Growth Rate Measurement: A Review

Peter C. McKeighan
Exponent Failure Analysis Associates, 4580 Weaver Parkway, Suite 100, Warrenville, IL, USA
pmckeighan@exponent.com

This chapter focuses on methods involved in, and issues to consider, when experimentally measuring the fatigue crack growth (FCG) rate behavior of a material. The scope will be wide ranging, touching on aspects of the standard method, as well as nonstandard, application-driven techniques.

One of the critical aspects of fatigue design is that the testing needs to be driven by the specific needs of the design approach, since fatigue testing is time consuming and expensive. The design approach utilized, and the level of detail and type of analyses employed, creates the need for what is required in the laboratory. Therefore, it is impossible to decouple the testing from the design process; the two are highly interdependent, as will be evident in the discussion below.

Materials Characterization: Modern Methods and Applications
Edited by Narayanaswami Ranganathan

ISBN 978-981-4613-06-4 (Hardcover), 978-981-4613-07-1 (eBook)
www.panstanford.com

7.1 Introduction

Conventional fatigue design is subdivided into two general areas, one utilizing a crack initiation approach (durability or safe life) as contrasted to one utilizing a crack propagation approach (damage tolerance). Both approaches are focused on reducing the structural weight, while ensuring that the desired fatigue design goals are met. However, the two approaches differ in their fundamental focus, with the crack initiation approach designing against crack initiation in a structure, while the crack propagation approach assumes the presence of a crack and, through calculation, shows that this crack will not grow to critical (failure) size before being detected during an inspection.

Structural design using either approach is highly dependent on having robust experimental data suitable for the loading environment and material in question. It is often the case that there is insufficient data available during design, which leads the design authority to perform testing relevant to the structure in question (this testing can take many forms and be at many levels, including basic material characterization, component or substructure testing spanning all the way to full-scale testing). In general, the experimental testing required to support a damage tolerant design is more involved than that based on crack initiation. Testing will, as a minimum, involve material property development (the basic crack growth rate of the material, as well as the fracture toughness at instability) as well as component-level testing under variable amplitude (VA) spectrum loading (the so-called coupon test phase). Some of the decisions involved with developing a damage tolerance (or fatigue crack propagation) test program to support design will be addressed in detail herein.

7.2 Preliminary Details: Developing a Testing Plan

One of the first key things to determine the scope of the fatigue test matrix of conditions will be to determine what is known, and unknown, regarding the specific material of interest. In this context, the material is not only the alloy utilized but also the product form (e.g., plate, sheet, extrusion, forging, etc.), the thickness, thermal processing (heat treatment or temper), and predominant grain orientation. Some materials, for instance, aluminum alloys 2024 and 7075, are well characterized in terms of the fatigue crack growth (FCG) rate. However, the level of characterization is highly dependent upon the product form and thickness.

Product form is a critical variable since this typically impacts both grain structure as well as residual stress state, both of which are primary drivers of fatigue performance. Therefore, FCG rate testing should be performed on the actual form of the product and, where possible, in the relevant thickness range of the product being designed. Ideally, the FCG rate specimens should be excised from the actual material, but this usually is not an option since design occurs well in advance of fabrication.

The scope of FCG testing required differs if you are validating existing data as opposed to developing new data. In the absence of relevant internal data from past designs, there are three excellent sources for FCG rate data: the *Damage Tolerant Design Handbook* [1], MMPDS [2] (previously known as MIL-HDBK-5), and the NASGRO database [3]. An example of a robust (wide-ranging in terms of stress ratio and extent of data) set of FCG rate data, namely da/dN versus ΔK,[a] is shown in Fig. 7.1 from Ref. [2] for 7075 aluminum.

The relevance of the test parameters in any pre-existing, directly applicable data should assist in defining the priority of subsequent testing to validate or fill in the gaps in the data.

The load ratio, R, is an important testing parameter during FCG evaluations. The load ratio is the ratio of minimum to maximum load (or stress) for a given fatigue cycle. It is often unnecessary and economically impractical to generate FCG rate data in the full range of positive and negative load ratios (an example of full range data is shown in Fig. 7.1). To determine what range is necessary, the predictive FCG model used in design needs to be considered,

[a] da/dN is the FCG rate defined by an incremental change in a crack of length a (da or Δa) divided by the cycle count, N, change required to achieve that increment (dN or ΔN). The stress intensity factor change, ΔK, can be viewed as a primary driver for crack growth and depends upon part geometry, crack geometry, stress level, and crack length. da/dN versus ΔK data is typically represented on a log-log plot with the linear region represented by a power law (the classic Paris regime of the data).

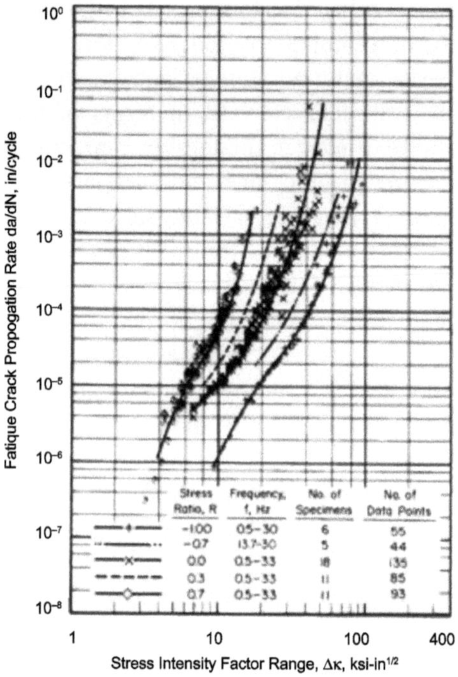

Figure 7.1 Example of FCG data contained within the MMPDS [22] for thin 2.3 mm 7075-T6 in the *L–T* orientation.

in particular how that model accounts for load ratio effects. In scoping out the testing required, at least the minimum amount of data to satisfy the parameters of the model(s) is recommended. To complicate matters, it is not uncommon for the growth rate response for positive and negative load ratios to be treated with different parameters. This issue is another example of a consideration required during the test-planning phase of a given test program to ensure that the data generated satisfies the needs of the design analysis.

The other issue that must be examined when determining the relevant load ratios for testing is the stress spectrum applied to the structure. For some ground-based structures, it is not uncommon for the majority of loading cycles to be at a fixed load ratio; in this case, FCG properties are necessary at only that load ratio. This simple situation is contrasted to aircraft design, where it is considerably

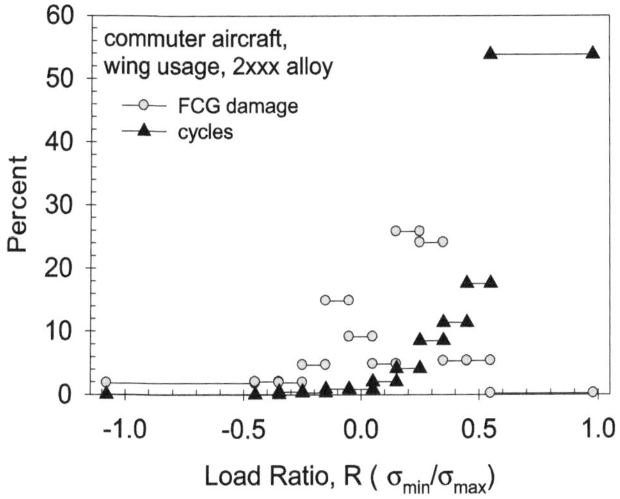

Figure 7.2 Example spectrum analysis illustrating damage content and cycle distribution as a function of load ratio *R*.

more complicated with different portions of the structure incurring sometimes very different cyclic loading conditions. In this case, it is quite helpful to examine the range of load ratios applied and, very importantly, the distribution of damage content for the cycling. The key is to let the testing be dictated by (a) the fatigue-critical locations that will be analyzed and (b) the expected structural loading (stress level and cyclic content) for each location.

There are a variety of methods available to assess damage content in the variable amplitude loading spectrum, but a simple approach that utilizes summed and normalized stress cycle raised to the Paris exponent (see Ref. [4] for more detail) is a reasonable first-order analysis approach. An example of this is shown in Fig. 7.2, where a wing spectrum for a commuter aircraft is examined. On the basis of purely the number of cycles, one might believe that the loading ratio range $R > 0.5$ is critically important since more than 50% of the applied cycles occur in this regime. However, as the plot in Fig. 7.2 shows, although a large number of cycles are applied at a high load ratio, these cycles account for less than 1% of the total damage. The vast majority of the damage (79% in total) occurs in the load ratio range of $R = -0.1$ to 0.3. Given this data, this would be the region where the FCG rate properties need to be quantified (and frankly, in this case the range is narrow enough that it might be sufficient to simply evaluate the FCG response at a single, low load ratio, perhaps $R = 0.1$).

The FCG rate curve, as graphed on a log-log scale, is typically subdivided into three regions (stages) depending upon the ΔK regime involved: (I) near threshold (lowest ΔK), (II) more linear, Paris-type regime, and (III) instability or near fracture [5]. In reality, the most economical, and hence the most frequently available, da/dN versus ΔK regions are stages II and III (a close examination of Fig. 7.1 indicates that this is the region of the data shown). The reason for this is that a near-threshold test in stage I can take significantly longer (weeks or months) to perform since measured growth rates are so slow. However, if a design stress spectrum includes numerous low-amplitude stress cycles, this regime could be more critical for fatigue. The typical design approach is to extrapolate stage II behavior to lower ΔK levels since this tends to yield somewhat conservative results (in other words, the extrapolation predicts more rapid crack growth than the asymptotic near-threshold-level data response). In a design situation where this degree of conservatism cannot be accommodated, it is important to test in the near-threshold regime.

Additional considerations regarding the environment should also be considered. Is this a high-temperature application? Is humidity or moisture an issue, or does the component operate in close proximity to seawater? Are there any other aggressive chemical species that may impact material performance? If the crack propagation process consists of a mechanical (cyclic) and time-dependent (e.g., corrosion or creep) component, this often means that the cyclic test frequency needs to be relatively slow. In more aggressive environments, the required cyclic rate can be very slow, hence creating practical challenges to performing testing in a timely manner. While these types of environmental variables add complexity to the fatigue testing required, they can prove critical to measuring the fatigue performance of the given material and as such will be discussed in more detail in a subsequent section.

In the absence of any corrosion or other time-dependent process, metallic materials can be cycled rapidly during testing, up to 100 cycles/second or sometimes even faster. However, lower-frequency limitations are typically applied to softer materials, for instance, polymers, where localized heating can take place and bias the results. Another example where testing frequency is limited is when testing joints with friction, since under cyclic loading conditions the heat generated by the faying surface contact cannot be dissipated fast enough.

7.3 Crack Length Measurement

In its simplest form, an FCG rate test consists of machining a specimen with a notch in it and cyclically loading the specimen (usually performed in a servohydraulic machine) so as to grow a crack from the notch, while periodically measuring crack length during the test. The key processed output from this test is the crack growth rate on a cyclic basis ($\mathrm{d}a/\mathrm{d}N$) and the stress intensity factor range (ΔK, derived from the load and specimen geometry) that drives the crack growth. Hence, a critical measured parameter during testing is the crack length in the specimen since this, along with applied cycles, is used to calculate the growth rate.

The most basic method for measuring crack length is via a travelling microscope, essentially a low-power microscope with a cross-hair reticule mounted on a vernier measurement scale, as shown in Fig. 7.3.

Landmarks are typically placed on the face of the specimen to provide reference points for relative measurement (the travelling microscope only provides relative measurement). Detecting a fine fatigue crack on the face of the specimen requires the surface to be polished to a near-mirror finish. Metallographic polishing wheels and portable buffers prove effective for this. Early implementations of travelling microscopes would typically also include a strobe light that pulsed with the loading cycle. By using a phase adjustment on the strobe, the crack could be viewed in "still mode" in the fully open state.

With the advent of higher-resolution digital cameras, more recent laboratory implementations allow continuous video imaging of the crack as it propagates across the specimen. This, then, allows

Figure 7.3 Travelling microscope viewing a cracked specimen.

the images to be postprocessed to yield incremental crack length values. When these systems are combined with powerful digital image correlation software, near-tip strains can be resolved around the crack tip to provide additional insight into the mechanism driving FCG.

Although there are a variety of nonvisual methods that can be used to measure crack length, all should be corrected (or essentially calibrated) to physical crack length measurements made on the specimen. An example of one method to correct nonvisual methods can be found in Ref. [6]. These corrections are especially important for thicker specimens (in excess of 6 mm) where crack front curvature can develop and not be reflected in the visual measurements made on the surface of the specimen.

7.3.1 *Compliance*

Inferring crack length from the compliance (defined as the reciprocal of stiffness) of the specimen is one of the oldest methods available for nonvisual crack length measurement. To utilize the compliance method, an extensometer is mounted to measure displacement on the specimen between two points on opposite sides of the crack as the specimen is loaded. What results is a load–displacement trace for that given loading cycle. The slope of the linear region of this data is calculated, and a closed-form solution suitable for the specific location on a given specimen geometry is used to calculate crack length.

Analysis of the load–displacement data to yield a compliance value is typically part of an automated testing package. However, the computer algorithm that processes the load–displacement data needs to be robust enough to cope with a variety of different issues. One example is if hysteresis is evident between the loading and unloading portion of the data. Oftentimes hysteresis is evident if there is significant friction in the gripping or antibuckling fixtures affixed to a given specimen. Yet another common issue is the presence of crack closure [7, 8], which results in nonlinear load–displacement data. Crack closure occurs as the crack "unzips" until fully open later in the loading cycle. This "unzipping" introduces an initial nonlinear region of the load–displacement behavior with linearity occurring only when the crack is fully open. Sometimes the load–displacement data can be used to determine the opening load of the crack. This can be an important quantity, given that any subsequent damage at the crack tip does not occur until the crack is fully open.

The optimum specimens to use for compliance are the edge cracked specimens where larger edge displacements increase the sensitivity of the method. The resolution of the technique is obviously less where displacements are smaller, for instance, with short crack testing. There are also resolution challenges with some specimen geometries; for instance, the middle cracked tension specimen (or surface crack specimen) where the overall stiffness of the specimen is relatively high and the sensitivity to a small stiffness change with a growing crack can sometimes be difficult to discern. If the gauge length of the displacement measurement equipment (essentially how far apart the blades of the extensometer are) is not much smaller than the crack of interest (in this context, smaller by a factor of 5 or more), sensitivity issues should be expected when applying the compliance method to discern crack length changes.

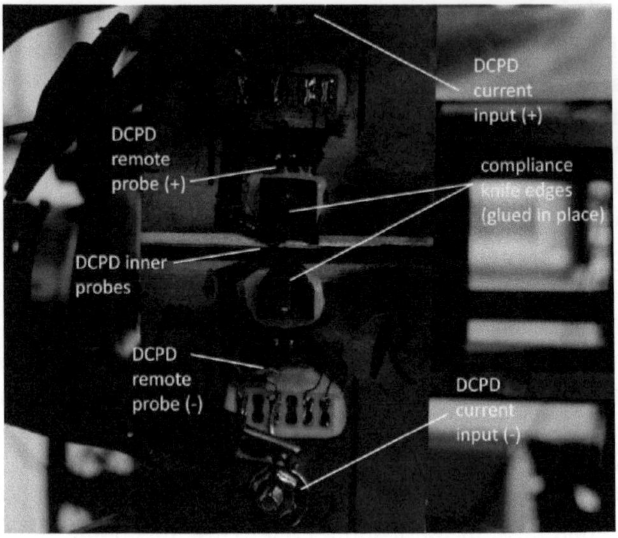

Figure 7.4 Center-cracked panel with compliance and DCPD instrumentation.

An example center-cracked panel specimen is shown in Fig. 7.4, with the knife edges for the extensometer glued onto the specimen (extensometry not shown for clarity).

These issues notwithstanding, the compliance method has been shown to be an extremely accurate technique for the measurement of crack length in fracture mechanics test specimens. The expressions used to derive crack length from compliance typically include an adjustable modulus parameter that is used to calibrate the method to visual measurements. In theory, this modulus parameter should be the same as the Young's modulus for the material; in practice, if larger differences are observed (e.g., a difference in excess of 10% of the Young's modulus), this is an indication that there may be an issue with the setup or compliance analysis techniques employed.

7.3.2 *DC Potential Drop*

Another commonly used technique to remotely (non-visually) monitor crack length during fatigue testing is the electrical potential, or potential drop (PD), method. With this technique, the direct current (DC) voltage disturbance in the potential field caused by a discontinuity (crack) is measured in a current-carrying body (either the specimen in a direct PD approach or a gauge affixed to the specimen in the indirect PD approach). As the crack grows, the effective cross-sectional area of the specimen decreases and the electrical resistance increases. Given a constant electrical current input, the magnitude of the voltage difference between two fixed points on either side of the crack increases and can be then related to the length of the crack [9].

Implementing this technique requires a highly sensitive and stable electronic system incorporating an accurate, constant current source and instrumentation capable of resolving the small voltage change that occurs between a given set of probes. Some commercial PD systems combine this instrumentation and provide an analog voltage output proportional to the change in voltage that can then be connected to an automated test system to measure crack length.

PD probe placement is critical for optimum accuracy. Typically, the desired position is marked with a scribe to facilitate placement. In practice, probes are spot-welded (if possible) on the face of the specimen with current input leads bolted through holes (due to the high current typically imparted to the specimen, on the order of 10s of amps). An example direct current potential drop (DCPD) specimen with current input screws and spot-welded probes is shown in Fig. 7.4. One of the reasons post-test correction is applied to correlate the physical measurements to the PD is that probes are inevitably mislocated to a minor degree.

One common PD implementation is to actually measure a voltage ratio change between dual sets of probes mounted on the specimen: a local set of probes close to the crack and a remote set of probes much further away from the crack. The advantage with this approach is that the stability requirements for the constant current source can be relaxed; variations are automatically accounted for when examining a voltage ratio. Furthermore, the setup is also impervious to temperature changes in the laboratory that can introduce significant offsets in the potential measurement, especially when crack growth rate test duration is long (weeks or months). Alternatively, this approach can be used for high-temperature crack growth rate testing as long as any temperature gradients in the specimen remain constant during the test. However, the dual-probe system does sacrifice some measurement sensitivity since the remote probes do typically register some voltage change as the crack grows. On balance, however, this sensitivity loss is a reasonable trade-off for the versatility and stability gained with the dual-probe system.

An additional advantage of the PD method is that when implemented in the direct manner (when the specimen is the electrical conductor), the method provides a through-thickness average of crack length. Hence, if any crack curvature or tunneling occurs, this is reflected in the measurement which then simplifies any post-test correction. Yet another benefit with the PD method is that calibrations are simple to develop; all that is needed for analog simulation is aluminum foil and a razor blade to create simulated cracks. This allows placement of the active, inner probes anywhere on the specimen. This is particularly valuable in the case of testing advanced materials where peculiarities in the crack path can occur. Examples of this include such issues as asymmetric crack growth in a center-cracked specimen [10], crack branching, and crack deflection [11]. As simple as the analog technique is, validation of calibrations that result from this method can be simply validated, or augmented with higher resolution, using an elementary heat conduction finite-element model.

The calibration issue (relating PD to crack length) is also more readily solved with precalibrated foil gauges that are fixed to the face of the specimen using the indirect implementation of the PD method. The disadvantage with this approach is that the crack lengths are surface crack lengths, not through-thickness-averaged. However, the method is well suited to testing nonconductive specimens. KRAK gauges (Hartrun Corporation, Eden Prairie, MN, USA) are also available for unusual crack growth geometries, such as the case of a crack growing out of a bolt hole.

7.3.3 *Other Methods*

There are a variety of other methods that have been used, with varying degrees of success, for nonvisual crack length determination. The preceding discussion has focused on the two most popular, nonvisual, fully automated methods to measure crack length. There is another PD technique—alternating current potential drop (ACPD)—that has been used successfully in the laboratory [12]. With ACPD, high-frequency currents of constant amplitude are passed through the specimen. These currents are self-confined in thin layers near the surface (skin effect) with the higher current density (compared to DCPD) better tailored to detect smaller cracks and nonpenetrating surface cracks. The degree of complexity and accuracy required in the ACPD electronic systems limits the ease of applicability (hence popularity of use) and, to some extent, the accuracy of these systems. Nevertheless, ACPD has proven effective in some applications, such as surface crack length measurement, and the method remains a niche technique for crack length determination.

Most of the other methods that are available and have been used for crack detection and sizing are hampered by the absence of a closed-form mathematical expression to relate field measurement to an actual crack size. For example, many of the techniques utilized for industrial nondestructive inspection have been used: ultrasonic, eddy current, radiography, and computerized tomography (CT) scan. These methods are particularly good for highly specialized, unique applications, but the absence of reliable calibration methods renders them unsuitable for widespread use.

One method, acoustic emission (AE), has been used quite successfully in detecting the presence of cracks as they nucleate and grow in a structure [13]. In this method, several microphones are placed on the specimen/structure at a number of locations, and as acoustic energy is released and moves through the structure, the data from the microphones can be used to triangulate the position of the source, in this case the damaging event. This method has been effectively applied to full-scale aging aircraft structural tests [14] and also to the hydrotesting of rail tank cars [15] to enable inspectors to focus efforts in detailed locations. However, as with the other methods, calibrating an acoustic energy signature to crack length has proven to be problematic, although there is some technical controversy concerning this point. Nevertheless, as a minimum, AE is a useful tool for detecting the earliest stage of damage formation and cracking.

7.4 Testing Standards

Although different organizations have developed their own highly specialized approach to FCG rate testing, the two publically available consensus standards are the ASTM E647 standard [16] and the International Organization for Standardization (ISO) 12108 test standard [17]. The two documents are fairly similar in content, with the ASTM predating the ISO standard by approximately 20 years. Although there is content in both standards that is unique, in general there tends to be more uniqueness in the longer ASTM standard.

There are several key differences between the standards that are important to note:

- When the loading cycle includes some compressive content, the ISO standard defines ΔK in a full range manner; for instance, if a cycle corresponds to endpoints of 15 and –5 $\text{MPa}\sqrt{\text{m}}$, then $\Delta K = 20\ \text{MPa}\sqrt{\text{m}}$.
- Conversely, using the same example as above, the ASTM standard includes only the positive portion of the cycle, so $\Delta K = 15\ \text{MPa}\sqrt{\text{m}}$.[a]
- Six specimen geometries are described in the ISO standard (compact tension, center-cracked tension, three types of bend specimens, and a single-edge notched tension specimen).
- Three specimen geometries are described in the ASTM standard (compact tension, middle-cracked tension, and the eccentrically loaded single-edge crack tension). However, it should be noted that the standard does allow any specimen geometry as long as a valid K-calibration is available

[a]These differences in ΔK definition when the loading cycle includes compression simply reflect the controversy in the technical field regarding relating fatigue crack growth to ΔK.

and the specimen is of sufficient planar size to remain predominantly elastic during testing.

- In the ASTM standard, there is provided a K-calibration, compliance calibration(s), and PD calibration(s) for each specimen described. Conversely, the currently accepted version of the ISO standard presents only a K-calibration, although efforts are currently underway in the standards development community to augment the ISO standard with compliance and DCPD calibrations.
- The ISO standard includes a single appendix focused on the PD (it was modeled after the similar ASTM/BSI material).
- The ASTM standard includes annexes/appendices on the following: PD, compliance, aqueous environments, data reduction, crack-opening load (closure), and small fatigue cracks.
- Finally, reflecting overall breadth, the ASTM standard cites 118 references in the technical literature and the ISO standard cites 49 references.

Whenever crack growth rate tests are performed, it is not uncommon to seek comparative data from the literature or from other sources. In this context, of the issues noted above, the one most critical is the ΔK definition for compressive cycling and how the data in the literature is represented.

7.5 FCG Testing Methods: da/dN versus ΔK

In previous sections, we have addressed some of the global test-planning issues, including how to choose conditions for testing, what instrumentation to use for crack growth measurement, and sources for additional test standard detail. The intent of this section is not to explore all of the intricacies of an FCG test in detail but rather to focus on some of the key decisions that are required during execution of a test. It is important to note that having a well-planned, simulated test—before actually starting testing or for that matter machining specimens—is the best insurance to meet program data requirements.

7.5.1 *Precracking*

Before the data-taking phase of the FCG test can start, a fatigue crack must be initiated from the machined notch and grown for a specific increment. Precracking is performed to ensure that the starting fatigue crack is sharp, the crack morphology (shape) is stable, and the effect of the machined notch is minimized. The precrack length required from the tip of the notch is typically fairly small, with a length equal to whichever of these quantities is larger: 10% of the thickness, the notch height/width, or 1 mm. The most effective starter notches are created by the electric discharge machine (EDM) process. EDM notches are quite narrow, typically less than 0.25 mm wide (even as low as 0.04 mm), with a brittle, recast layer that readily nucleates cracks under cyclic loading.

In general, the load levels applied during precracking must be less than those applied during the portion of the test when the crack growth rate is being measured. However, the test standards do allow precracking at higher load levels, provided a load shedding (reduction) method is employed during precracking. Care must be taken to ensure that the precrack load history does not have an effect on the crack growth rate measured during the test. Guidelines to this effect have been included in the FCG test standards.

There has been a fair bit of recent interest in performing precracking under fully compressive loading cycles. Although this is a controversial issue, one segment of the technical community believes that this is important when generating near-threshold FCG rate data [18]. Efforts are currently underway within the standards development community of the ASTM to draft a document outlining how this methodology should be performed. This document will presumably be part of the ASTM E647 FCG rate standard [16] soon so as to harmonize the ongoing testing utilizing the technique.

7.5.2 *Load Control during FCG Testing*

In the early days of the ASTM FCG standard, the vast majority of the FCG data was produced under constant amplitude loading conditions. Using this approach, the ΔK range for a given test was dictated by the natural K-gradient of the specimen (the K-gradient is the rate at which K changes with crack length), the length of the precrack, and the plasticity limits on the uncracked ligament. Some specimens, for instance, the middle-cracked tension panel, exhibit a fairly shallow K-gradient. Therefore, to measure the full range of crack growth rate data, under constant-amplitude loading conditions, two separate specimens typically need to be tested, which is expensive and time consuming. Furthermore, the constant-amplitude test procedure is well suited for crack growth rates in excess of 10^{-8} m/cycle; below this level, precracking considerations increase the difficulty and time required to test.

Other K-control specimen loading options are now available with the advent of more sophisticated test control systems, faster computers, and enhanced resolution instrumentation. However, there remains some controversy in the technical community regarding the methods and potential load history effects on the resulting FCG rate data. One of the key parameters in K-control testing is the normalized K-gradient, C; this parameter is defined by $C = 1/K\ \mathrm{d}K/\mathrm{d}a$. Possible K-control methods include the following:

- K-decreasing methods: suitable for generating $\mathrm{d}a/\mathrm{d}N$ versus ΔK data at near-threshold levels by moving down the curve to lower ΔK levels
- K-increasing methods: more rapid generation of $\mathrm{d}a/\mathrm{d}N$ versus ΔK data for the fullest range possible on the specimen
- Constant ΔK methods (which results in a constant growth rate, the utility of which will be discussed later)
- Constant K-max methods: a variable load ratio approach with either increasing or decreasing K-min (the increasing K-min approach allowing generation of a conservative threshold crack growth rate at a high load ratio)

The specific parameters and limits for application of these methods, while beyond the scope of this document, are clearly denoted in both the ISO and ASTM specifications [16, 17]. However, before any of these advanced K-control methods are undertaken, efforts should be made to assume a crack growth rate curve and simulate the test on paper before testing starts in order to ensure that the desired range of growth rate and ΔK are achieved.

Although the ASTM test standard provides guidelines with regard to acceptable values of the normalized K-gradient C, the following note is provided [16]:

> Note 8: Acceptable values of C may depend on load ratio, test material, and environment. Values of C algebraically greater [than the limit] indicated have been demonstrated as acceptable for us in decreasing K tests of several steel alloys and aluminum alloys tested in laboratory air over a wide range of force ratios.

As another example, it has generally been established that when FCG testing high strength titanium, C values can be extended to well beyond the limit contained within the ASTM standard. Knowing this would be critical for planning, testing, and sizing specimens, although extreme care must be taken when testing outside the accepted guidelines of the FCG test standards. In summary, it is always useful to perform a material-relevant literature search before testing/planning commences to assist in selecting the conditions and methods used during FCG rate testing.

7.5.3 *Crack Growth Rate Data Analysis*

An FCG test yields crack length as a function of cycle count, as shown in the typical data in Fig. 7.5, with the FCG rate shown in Fig. 7.6 calculated from the instantaneous slope of the a versus N data. In Appendix X1, the ASTM E647 [16] test standard, two data analysis methods are presented to calculate $\mathrm{d}a/\mathrm{d}N$ from crack length versus cycle count data. The key difference between these two methods is the extent of data smoothing that occurs in calculating the $\mathrm{d}a/\mathrm{d}N$ slope.

The first method is the secant method, whereby the growth rate at a given point in Fig. 7.5 is simply the slope of the line connecting the previous and following data points. This is a simple and robust approach for growth rate calculation, essentially smoothing between three data points.

The second $\mathrm{d}a/\mathrm{d}N$ analysis method, presented in computer program form in Appendix X1 of ASTM E647 [16], performs a polynomial fit to seven points and calculates the slope at the middle

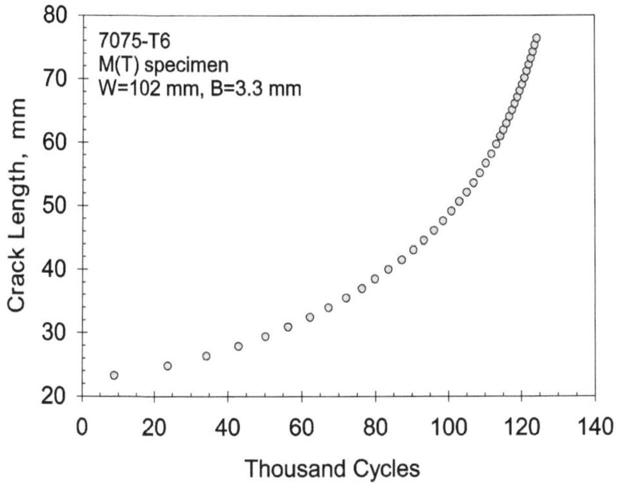

Figure 7.5 Typical crack length versus cycle count data for an FCG test.

fourth point. Hence, in this case the smoothing occurs over seven individual data points.

When analyzing crack length and cycle count data, a critical issue that has to be carefully controlled is the crack length interval between data points, Δa. Modern automated control systems can provide much more data than necessary due to the resolution and speed of these systems. However, the FCG rate standards provide clear direction for how the crack length versus cycle count data need to be spaced with regard to crack length interval, Δa. For instance, the data shown in Fig. 7.5, which conforms to the guidelines of E647, corresponds to approximately every 10th data point of the total recorded by the test control system. Hence, the Δa used in calculating the FCG rate is 10 times the control system measurement precision.

Parsing the data in this manner is essential to reduce the corresponding noise level of the growth rate calculation. This difference is graphically shown in Fig. 7.6, where the circular data points (fine analysis) treat every recorded point and the triangles (coarse analysis) space the data in accordance with the ASTM E647 guidelines. Note that these two analysis methods were applied to the same basic set of recorded a versus N data. Clearly, at a fixed ΔK

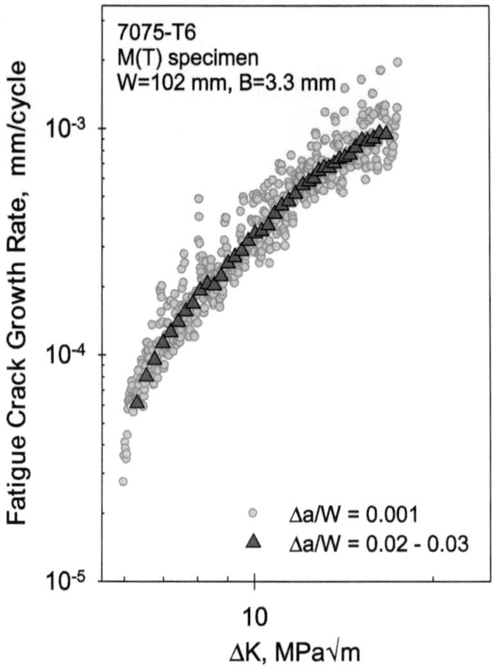

Figure 7.6 Influence of crack length interval on the scatter in the FCG rate.

level, the noise band in the data appears to be approximately a factor of 2× to 3× on the FCG rate. Conversely, the coarse analysis FCG rate is considerably smoothed and more in accordance with expectation.

7.5.4 *Fatigue Crack Growth Rate Variability*

The observed variability in the crack growth rate at a given ΔK is important to consider for comparative purposes once data has been generated in a given FCG evaluation. This variability issue also has implications from the viewpoint of how the data is used in a design context and the perceived accuracy of any life prediction that may result.

This variability issue has been examined extensively in studies supporting the ASTM FCG test standard [16]. A recent interlaboratory test program included 18 labs and 3 materials (4130 steel, 7075 aluminum, and 2024 aluminum) with a scope of 141

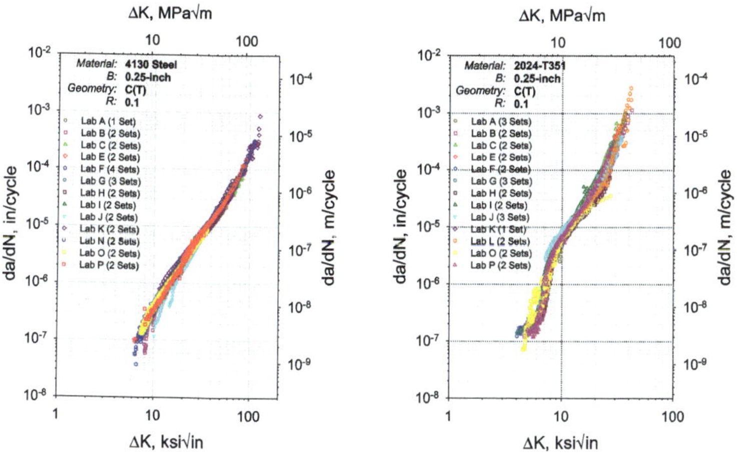

Figure 7.7 Recent FCG round-robin variability results for 4130 steel and 2024 aluminum at $R = 0.1$ for a thickness of 6.35 mm [19].

individual tests [19]. Interlaboratory (lab-to-lab) variability was lower for the steel (da/dN variability of $\pm$ 31%, or a factor of 1.9$\times$) than for the aluminum materials (da/dN variability of $\pm$ 41%, or a factor of 2.4$\times$). In general, the rough rule of thumb is that interlaboratory variability is a factor of 2$\times$ on da/dN with intralaboratory (replication within a lab) variability less by about 1/3. To represent some of these variability numbers in a graphical context, a limited amount of the data from Ref. [19] is indicated in Fig. 7.7 for steel and aluminum at a low load ratio.

Data variability is an important consideration when planning an FCG test program. Experience has shown that all unique test conditions should be evaluated with at least two tests to replicate behavior when testing in stage II Paris regime. In practice, planning for 2 or 3 tests per condition (machine specimens sufficient for 2.5 tests, meaning 2 or 3 tests are performed with equal frequency) is safest to accommodate occasional unusual behavior and the inevitable lack of replication. This approach is contrasted to a higher level of replication required when testing in the near-threshold region. In this case, a minimum of three or four specimens should be planned for each evaluation.

7.6 FCG Testing Methods: Variable Load Amplitude

Whereas the discussion so far has focused on generating basic material data (da/dN versus ΔK), crack growth experiments under VA spectrum loading conditions provide empirical validation of the fatigue design for structures subjected to highly varying conditions [20]. When the load amplitude applied to a structure varies over time, depending upon usage, this loading variation can cause either retardation (usual) or acceleration (atypical, more related to highly compressive loading) in crack growth behavior. The impact of these load interaction effects on fatigue life can be significant; factors of $3\times$ or more are not uncommon. Hence, not accounting for these effects can lead to large, unnecessary maintenance costs (needless inspections) over the lifetime of a structure or, in the rarer case of crack acceleration, nonconservative fatigue life prediction.

Unfortunately, the complexity of a given VA spectrum makes "blind" analytical prediction of the resulting impact on fatigue life problematic. Therefore, the load interaction models available in the fatigue design process require experimental input to tune the empirical parameters in the models. In practice, this leads to experimental crack growth testing under spectrum loading conditions to support the damage tolerant design process [21].

7.6.1 *Essence of a Spectrum Crack Growth Coupon Test*

An FCG test under VA spectrum loading conditions (spectrum crack growth, or SCG, test) can be considered as a highly specialized FCG test. Although to date, no standardization has ever been attempted with this type of test [22], many of the features of a standard FCG test apply with the following exceptions:

- The loading corresponds to a discrete number of highly VA cycles that are continually repeated during the test.
- Although the test specimen (called a coupon) can be a standard configuration, it is commonly not, often encompassing detailed geometric features characteristic of the modeled structure.

- When the coupon is not a standard configuration and it exhibits a thickness change or some complicated geometric detail, visual crack length measurement is typically relied upon since the absence of calibration makes it difficult to apply any standard nonvisual measurement approach (e.g., compliance or PD).
- The crack geometry in an SCG test is rarely a through crack; more typically, it is a corner crack near a geometric feature or, in some cases, a surface crack.
- The output from the test is typically the crack length versus cycle count data that a designer would then use to empirically calibrate a load interaction model.

In general, the uniqueness of a given SCG test implies that it is more of a structural test than a materials test, since the results are largely dependent upon the geometry of the specimen and the VA loading applied. Although this should be obvious, the key to ensure applicability of the experimental program is for the test engineer to work closely with the design engineer.

7.6.2 *Test Program Detail*

Since the fracture mechanics models used in design tend to be fairly simplified, it is not uncommon to have a structural detail where the geometry is not directly applicable to the simplified model (e.g., a stress concentration compounded with a thickness change). In this case, the salient geometric feature(s) in the structure is/are captured in the coupon to include geometric effects combined with the influence of the VA loading. For reference, some examples of SCG test coupons are included in Fig. 7.8. These examples range from simple plate specimens with local geometry changes to more complicated joints that include fastener filled holes and varying load transfer.

Until this point, the cracks discussed in the course of FCG rate testing have been one-dimensional cracks; in other words, they can be characterized by a single crack length measurement (on the front and back of the specimen). During SCG testing, it is typical to test two-dimensional cracks, for instance, corner cracks

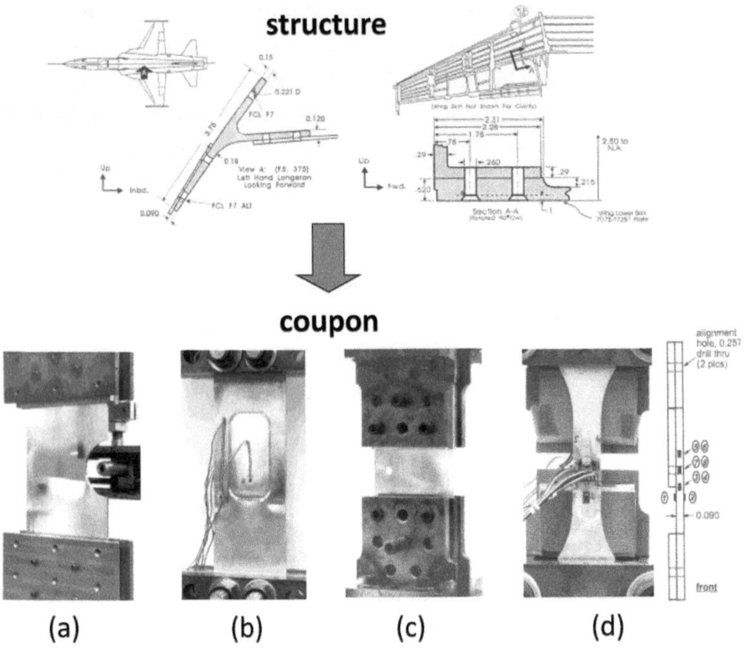

Figure 7.8 Test coupons reflect key features of structure detail for (a) an edge crack at a radius, (b) a surface crack near a scallop detail, (c) cracking from the edge of a fastener hole, and (d) multiple-layer joint simulation with low levels of load transfer [15].

or surface cracks, where at least two crack length measurements are required to quantify crack shape. Notches suitable for cracking are usually formed in the specimen with either sinker (plunging) or wire electrodischarge machining. Electrodischarge machining is typically used to create the starter notch for the following reason: rapid initiation is desired and the luxury to grow a long precrack is usually not an option since growth rate data is usually required before the crack is 1.27 mm long. Where smaller notches/cracks are required (e.g., when testing turbine engine components), a focused ion beam technique can be utilized to create a notch.

The notch/crack geometry required for a given test is dictated by the geometry of the fatigue critical location being analyzed. In the case of corner cracks, a simple triangular cut at a free edge will rapidly transition to a curved crack during precracking. Sometimes

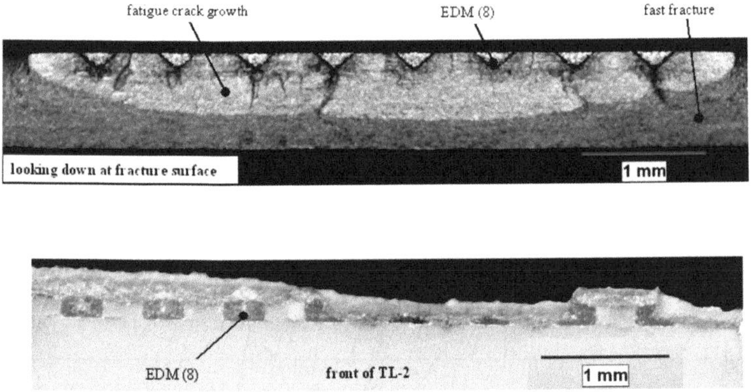

Figure 7.9 Example of a long-on-the-surface, shallow-depth crack geometry.

the crack shape is of particular interest for a design. An example of this is shown in Fig. 7.9, where a linear series of small triangular-shaped EDM plunges (divots) resulted in a high-aspect-ratio crack (long on the surface and shallow in depth).

During precracking, cracks nucleated at each divot, coalesced with neighboring cracks, and formed the overall high-aspect-ratio crack.

The magnitude of the peak load during precracking is typically controlled as a percentage of the peak load in the spectrum. An approximate rule of thumb is to start precracking at a constant-amplitude low load ratio with the maximum load 50% of the peak spectrum load. Higher levels are sometimes necessary for more benign spectra. It is believed that crack wake effects during spectrum precracking are not an issue once at least one pass through the spectrum has occurred.

The process to develop a load spectrum suitable for testing is beyond the scope of this discussion. Suffice to say that multiple treatises could be written addressing the issue of how spectra are measured, processed, and developed. Spectrum editing issues—for instance, ordering, clipping, and truncation—are often issues that need some type of empirical validation. Yet another issue is spectrum scaling. The test spectra (also referred to as a "usage" or "sequence" when describing a particular mission mix or series of maneuvers) are often common between a given series of fatigue-critical locations with only different scale factors applied to the spectrum for the different locations. Be aware that the scale factor (essentially proportional to stress) can make a large difference in the resulting fatigue performance in the coupon.

As indicated before, the primary result of an SCG rate test is the crack length as a function of cycles. The criteria typically applied to determine whether the given two tests need to be repeated is based on the cyclic life for a given crack length interval (usually 80%–90% of the full range measured during the test, starting with as small a crack length as possible with the test data). If the cycles required for this crack length interval are within ± 10%–15% from each other, the test is deemed repeatable and a replicate is unnecessary [22].

7.7 Other Miscellaneous Issues

The applications discussed so far have tended to be the typical testing situation called for in the vast majority of applications for which FCG testing is required. Occasionally a situation arises where more highly specialized capabilities are necessary as will be discussed below.

7.7.1 *Testing in an Environment*

The most common nonambient environment for FCG testing is either high humid air (HHA) or an aqueous water environment (the ASTM test standard has an annex with guidelines for testing in aqueous environments). HHA conditions are easily simulated in the laboratory with a plastic bag cut to accommodate the specimen and taped flush to the specimen surface with duct tape. A saturated sponge placed in the bag, provided it is well sealed, will create relative humidity levels typically 95% and higher. The plastic bag makes periodic visual crack length measurement problematic, but as long as visual measurements are made before and near the end of the test, this is not an issue. The PD is ideal for this situation since compliance requires an extensometer that might not be amenable to

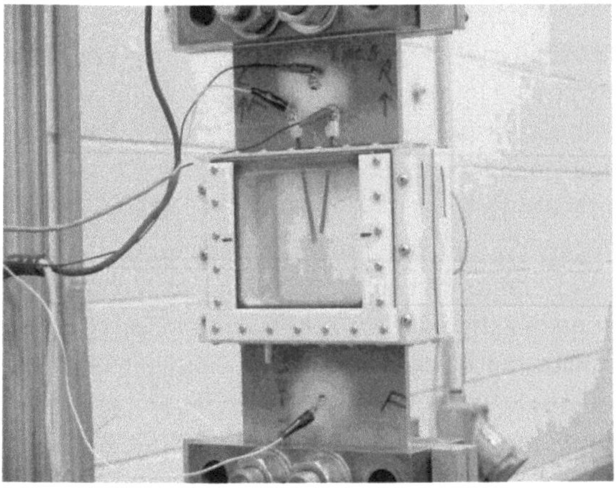

Figure 7.10 Fatigue crack growth testing in JP-8 jet fuel.

the environment. However, care must be taken since condensation on the crack face can lead to noisy or erroneous PD readings.

The aforementioned approach with the plastic bag can be used with varying degrees of success for fully aqueous environments. The more typical approach is to fabricate a chamber that mounts flush to the specimen in order to fully wet the crack. An example of such a chamber is shown in Fig. 7.10 for an application where the environment was JP-8 jet fuel. This approach offers the advantage of being able to measure crack length, provided dimensional landmarks are available on the specimen to make relative measurements with a travelling microscope. Sometimes it is convenient to rotate the test machine 90 degrees in a horizontal configuration to allow the ligament of the specimen to be submerged and the edge of the specimen exposed to air (e.g., to accommodate compliance measurement).

A similar bagged arrangement has been used with special pin seals on the compact tension specimen. A schematic of this sealing arrangement is available in Ref. [23]. Alternative approaches have been used with more complicated chambers for sealing compact tension specimens. These approaches have been effective for containing even the most aggressive environments. This containment is most easily achieved by having the clevis grips in the chamber and sealing on the cylindrical rods that mate to the top of the grips. The primary challenge with this approach is that the seals to the loading rods must be able to allow the loading rods to slip as the specimen is cycled. It is also imperative to have clevis grips that are impervious to the environment contained within.

One of the most critical variables when testing in an environment is the frequency, and potentially the shape, of the loading cycle. Frequency is important since the damage in the specimen is a combination of a time-dependent environmental component superimposed with the mechanical (cyclic) fatigue loading component. The general rule of thumb is that when testing in HHA or aqueous water solutions, the frequency should be no higher than 1 Hz. A literature search is strongly recommended before undertaking testing in an environment where the experimenter's first-hand experience is limited.

In more aggressive environments, frequencies even lower than 1 Hz may be required to ensure that the environmental damage component is fully activated. An example of this is shown in Fig. 7.11 for FCG testing on steel in an industrial proprietary, aggressive, chemical environment. This data was measured by performing constant ΔK FCG tests (a constant growth rate test) in the environment, while varying the cyclic frequency. Although the data is mixed, the trend is of increasing damage content with decreasing frequency, even down to the lowest 0.002 Hz (one would expect the ratio to plateau at and below the frequency when the full environmental effect is achieved—the fastest-possible test frequency). This environment is a particularly difficult one in which to perform cyclic testing in any practical amount of time. In a situation such as this, the only choices available are to (1) perform testing at a more rapid frequency but use data such as that shown in Fig. 7.11 to "correct" the growth rate data that is measured or (2) test at a phenomenological and design-relevant frequency.

7.7.2 *Testing with Residual Stresses*

Residual stress has always been important for fatigue, as evidenced in the earliest development of peening in the rail industry to mitigate fatigue failure. However, its primary impact is on fatigue crack

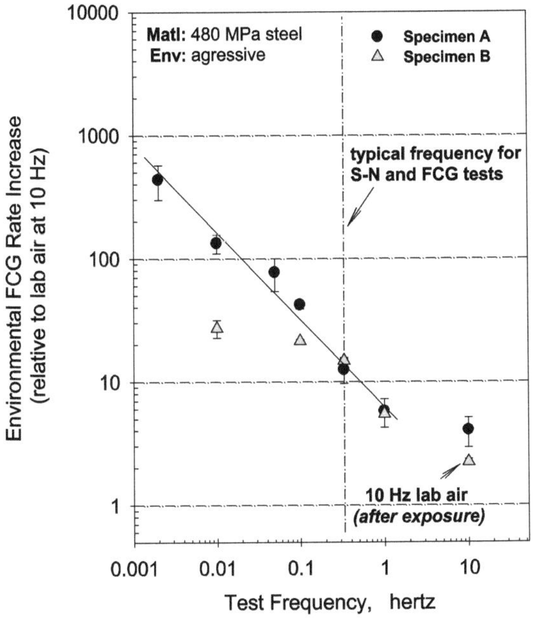

Figure 7.11 Observed frequency effect at constant ΔK conditions in an aggressive, industrial proprietary environment [14].

initiation, and only recently has there been more interest in the impact on crack propagation. A tensile residual stress will impact FCG by increasing the "effective" load ratio to which the material is subjected since it acts to increase the mean stress in the material.

The resurgence of late in residual stress effects on fatigue crack propagation has been spurred by evolving fatigue life extension methods (e.g., cold expansion of fastener holes and laser peening), as well as recent application of thick section aluminum forgings. A paper by Ball [24] proposes a new damage-tolerant design methodology that accounts for the presence of residual stresses. The traditional approach, using lower-bound FCG properties, was adding undue conservatism in the design. One key aspect of this approach is analyzing FCG rate data using an adjusted compliance ratio [25] that has been shown to account for residual stress. Effort is currently underway within ASTM to standardize this method to measure adjusted compliance ratio during FCG rate tests.

One FCG test method that can be used to indirectly discern residual stress effects is the constant ΔK test discussed earlier in the context of environmental testing. In this test, possible with the most sophisticated FCG test control systems now available, a constant crack growth rate is observed. However, as the crack grows through residual stresses, the growth rate is perturbed, and knowing how FCG rate varies with load ratio in the material, the residual stress can then be inferred.

One of the challenges when testing a section that has a residual stress is that when a specimen geometry is cut out of the part, the local residual stress is relieved as the piece is excised. This is most apparent when a piece of thicker plate material is fly-cut-milled into a thinner FCG sample. If the machining is first performed on one side of the plate and it is then flipped to machine the opposite side the same amount, the piece, when released from the machining chuck, is warped due to residual stress relaxation. The solution is to make many small machining passes removing a minimal amount of material and flipping the stock between each pass. For the same reason, it is often desirable to test as large a section as possible to preserve the residual stresses in an FCG specimen.

How to accommodate residual stress in design remains a controversial issue. Within the military aerospace community in the U.S., residual stresses are noted and measured but generally not accounted for in design. One of the reasons for this is the wide variability that can occur in the magnitude of the residual stresses. Moreover, the techniques for full-field, three-dimensional strain measurement are neither well suited for industrial application nor available in the size necessary for aircraft structure. Part-to-part and processing variability has also been observed in the past to have an impact on residual stress magnitude.

7.7.3 *Testing with Welds*

Fracture testing in welds is common, although occasionally the interest is in crack growth properties as well. Many of the aforementioned issues with regard to residual stresses are present when testing welds. The two areas in and around a weld that are of most interest are directly in the weld metal and adjacent to the weld in the heat-affected zone (HAZ). The HAZ is in the region where the highest tensile residual stresses are present and hence its propensity to crack is elevated. When testing in the HAZ, a crack starter is located approximately 1 mm from the edge of the weld (cross-sectioning the welds in several places and etching the microstructure to observe heat affected areas can be useful in determining where the HAZ is) and precracked. Often times, the crack will meander to the HAZ and run along it naturally since this is the "fatigue weak" location.

Microstructural and strength differences in and around welds can affect FCG rate as well. Testing in the weld metal is typically an easier prospect with the crack starter notch in the middle of the weld. It is not uncommon, however, for the crack to exhibit tortuous growth with numerous deflections and bifurcations due to the highly variable microstructure. Nor is it uncommon for the crack to grow out of the weld metal and into the HAZ. In either case, it is judicious to plan on taking more visual crack length measurements than might typically be done in parent metal.

The FCG rate data generated in and around welds tends to exhibit a higher degree of variability than parent material. A higher degree of replicate data should be planned for in and around welds (at least twice as many tests as in well-behaved parent material). Visual crack length measurements will also likely require that a portion of the weld be ground and carefully polished.

References

1. Skinn, D.A., Gallagher, J.P., Berens, A.P., Huber, P.D., and Smith, J. (1994). *Damage Tolerant Design Handbook*, WL-TR-94-4052 (Materials Directorate, Wright Laboratory, Wright Patterson AFB).
2. Rice, R.R., Jackson, J.L., Bakuckas, J., and Thompson, S. (2003). *Metallic Materials Properties Development and Standardization (MMPDS)*, DOT/FAA/AR-MMPDS-01.
3. NASA-JSC and SwRI (2012). NASGRO Fracture Mechanics and Fatigue Crack Growth Analysis Software.
4. McKeighan, P.C., Fess II, F., Petit, M., and Campbell, F.S. (2000). Quantifying the magnitude and effect of loading errors during fatigue

crack growth testing under constant and variable amplitude loading, in *Applications of Automation Technology in Fatigue and Fracture Testing and Analysis: Fourth Volume*, ASTM STP 1411, Braun, A. A., McKeighan, P.C., Nicolson, A.M., and Lohr, R.D., eds. (ASTM International, West Conshohocken, PA).

5. Suresh, S. (1998). *Fatigue of Materials*, 2nd Ed. (Cambridge University Press).
6. McKeighan, P.C., and Smith, D.J. (1994). Determining the potential drop calibration of a fatigue crack growth specimen subject to limited experimental observations, *J. Test. Eval.*, **22**(4), 291–301.
7. Elber, W. (1970). Fatigue crack closure under cyclic tension, *Eng. Fract. Mech.*, **2**, 37–45.
8. Elber, W. (1971). The significance of fatigue crack closure, in *Damage Tolerance in Aircraft Structures*, ASTM STP 486 (ASTM International, West Conshohocken, PA), 230–242.
9. Johnson, H.H. (1965). Calibrating the electrical potential method for studying slow crack growth, *Mater. Res. Standards*, **5**(9), 442–445.
10. McKeighan, P.C., Zhu, W-X., and Smith, D.J. (1995). A full field DC potential drop calibration for an asymmetrically cracked M(T) specimen, *J. Test. Eval.*, 102–110.
11. McKeighan, P.C., Tabrett, C.P., and Smith, D.J. (1995). The influence of crack deflection and bifurcation on DC potential drop calibration, in *Special Applications and Advanced Techniques for Crack Size Determination*, ASTM STP 1251, Ruschau, J.J., and Donald, J.K., eds. (ASTM International, West Conshohocken, PA), 51–66.
12. Ruschau, K.K., and Donald, J.K., eds. (1995). *Special Applications and Advanced Techniques for Crack Size Determination*, ASTM STP 1251 (ASTM International, West Conshohocken, PA).
13. McKeighan, P.C., Perocchi, L.C., Nicholls, A.E., and McClung, R.C. (1999). Characterizing the cracking behavior of hard alpha defects in rotor grade Ti-6-4 alloy, in *Fatigue Behavior of Titanium Alloys*, Boyer, R.R., Eylon, D., and Lutjering, G., eds. (Minerals, Metals and Materials Society, Rosemont, IL), 349–356.
14. McBride, S.L., Viner, M.R., and Pollard, M.D. (1991). Acoustic emission monitoring of aging aircraft structures, in *Review of Progress in Quantitative Nondestructive Evaluation*, **11B**, Proceedings of the 18th Annual Review, Brunswick, ME, 2275–2282.
15. McBride, S.L. (1994). *Acoustic Emission Tank Car Test Method Review and Evaluation*, **93**(6), Canadian Institute of Guided Group Transport,

Transportation Development Centre–Policy and Coordination Group, Queen's University.

16. ASTM E647-11 (2012). Standard test method for measurement of fatigue crack growth rates, *2012 Annual Book of ASTM Standards*, Section 3, Metals Test Methods and Analytical Procedures, 675–720.
17. International Standard ISO 12108 (2012). *Metallic Materials: Fatigue Testing; Fatigue Crack Growth Method*, 2nd Ed., 1–43.
18. Newman, Jr., J.C., and Yamada, Y. (2010). Compression precracking methods to generate near-threshold fatigue-crack growth-rate data, *Int. J. Fatigue*, **32**, 879–885.
19. McKeighan, P.C., Feiger, J.H., and McKnight, D.H. (2008). Round robin test program and results for fatigue crack growth measurement in support of ASTM standard E647, *ASTM Research Report E08-1007* (ASTM International, Committee E08 on Fatigue and Fracture, Subcommittee E08.06 on Crack Growth Behavior, West Conshohocken, PA).
20. Broek, D. (1989). *The Practical Use of Fracture Mechanics* (Kluwer Academic).
21. McKeighan, P.C., Garcia, D.B., Cardinal, J.W., and Thomsen, M.L. (2006). The role of mechanical testing in the damage tolerance process, in *Proceedings of the SF2M International Conference on Residual Fatigue Life and Life Extension of In-Service Structures*, JIP 2006, 11th International Spring Meeting, Paris, France, 71–82.
22. McKeighan, P.C., and McMaster, F.J. (2003). A framework for a standardization effort for fatigue crack growth testing under variable amplitude spectrum loading, in *Fatigue Testing and Analysis Under Variable Amplitude Loading Conditions*, ASTM STP 1439, McKeighan, P.C., and Ranganathan, N., eds. (ASTM International, West Conshohocken, PA).
23. McKeighan, P.C. (2002). Testing of metallic materials, in *Handbook of Materials Selection*, Kutz, M., ed. (John Wiley and Sons), 519–543.
24. Ball, D.L. (2009). The influence of residual stress on the design of aircraft primary structure, in *Fatigue and Fracture Mechanics: 36th Volume*, ASTM STP 1508, Neu, R. W., Wallin, K.R.W., and Thompson, S.R., eds. (ASTM International, West Conshohocken, PA), 216–239.
25. Donald, J.K., Bray, G.H., and Bush, R.W. (1998). An evaluation of the adjusted compliance ratio technique for determining the effective stress intensity factor, in *29th National Symposium on Fatigue and Fracture Mechanics*, ASTM STP 1332, Panontin, T.L., and Sheppard, S.D., eds. (ASTM International, West Conshohocken, PA).

Chapter 8

Measurement of Wear and Friction Resistance of Bulk and Coated Materials

Caroline Richard,[a] Geetha Manivasagam,[b] and Yan Ming Chen[c]

[a]*Laboratoire de Mécanique et de Rhéologie (LMR) EA2640,*
Université François Rabelais de Tours (UFRT), Polytech 'Tours,
7 avenue Marcel Dassault, Tours, 3700, France
[b]*Centre for Biomaterials Science and Technology,*
School of Mechanical and Building Sciences, VIT University,
Vellore, 632 014, TN, India
[c]*Centre Technique des Industries Mécaniques (Cetim), Laboratoire de Tribologie,*
Senlis, France
caroline.richard@univ-tours.fr

8.1 Introduction

To increase the reliability of parts in friction and the life cycle of tools, engineers often propose solutions by optimization of their design, the choice of materials, and lubrication and surface treatments. The security and economic stake of machines often requires a preliminary validation by testing in the laboratory before integrating new solutions into the system, in particular in the field of transport and power production. The choice of a tribological test method representative of the real system thus becomes a stage of first magnitude in the success of the whole project.

Materials Characterization: Modern Methods and Applications
Edited by Narayanaswami Ranganathan

ISBN 978-981-4613-06-4 (Hardcover), 978-981-4613-07-1 (eBook)
www.panstanford.com

The objective of this chapter is to provide engineers and technicians with a global and structured vision of tribological tests existing from the classification of tests according to the needs of the industry. The presentation of steps of choice and the techniques of measurement are detailed through typical examples of tribometers. A summary table of the current tribological tests gathers a great majority of the tests according to the type of contact or configuration.

8.2 Elements of Tribological Testing

8.2.1 *Economic Importance and Technological Particularities*

Tribology is defined as the sciences and technologies that investigate the interactions between two surfaces in contact. It includes three domains: friction, wear and lubrication. Even if the introduction of the word "tribology" by Peter Joste in his report published during the oil crisis of the 1970s does not correspond to the birth of this science, which began with the history of humanity, the development of tribology is always strongly connected to the economic situation, in particular since the Industrial Revolution. The results of a study published by Cetim in 1994 confirm once again that the potentially practicable economy is of the order of 26 billion euros a year in France, that, is 1% of the gross domestic product (GDP), if all the knowledge acquired in tribology were applied to industry [1].

If tribology plays a significant role in economic development, tribometry or tribological tests constitute an essential basis of this science because the physical knowledge on numerous phenomena of contact still remains to deepen, in particular concerning adhesion, fretting, or cavitation. Most of the laws in tribology (Coulomb, Achard) still result from experiments in specific domains.

Tribometry is defined as a set of technologies that allows measuring of the tribological behavior of a system in friction. In comparison with the measures of the mechanical resistance of materials, which depends on the cohesion of its components, the measure of tribological behavior is sharply more complex because

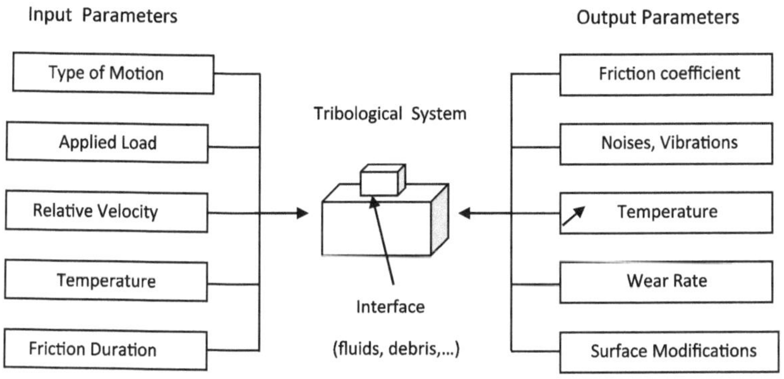

Figure 8.1 Tribosystem.

of the nature of this behavior, which is not an intrinsic property but an answer of a system.

In the mechanical engineering industry, tribological tests are indispensable in numerous stages of production and manufacturing. Tribology is utilized, starting from metal working, such as rolling, forging, forming, etc.; manufacturing of mechanical components like brakes of vehicles; commands of flights of planes; and cash dispensers, to control friction and guarantee the life cycle by optimizing the choice of materials and surface treatments in order to ensure the quality and reliability of states of surface and lubrication.

8.2.2 *Objectives and Classification of Tribological Tests*

As the tribological behavior is not an intrinsic property of materials, but an answer of a tribosystem (Fig. 8.1), it is necessary to know this system before carrying out any measurement.

A tribological system includes four elements [2]:

- Two parts (first bodies) in contact (materials, geometrical configuration, surface roughness)
- A third body (lubricant, wear debris, transfer layers, pollutant)
- The environment (temperature, humidity)

With input parameters exercised on both parts in contact, pressures of contact, impact or sliding velocities, etc., these four elements give answers resulting from interactions. The objective of tribological tests is to identify answers of this system (friction force, wear, acoustic emissions and vibrations, heating) according to the given input data (pressure of contact, sliding speed, temperature).

Tribological tests can be classified according to fields of application and modes of wear and in the first case, they can be classified into three categories:

- **Fundamental tests:** Fundamental tests aim at a deeper understanding of the basic phenomena of tribology. The design of these tests is often simple and ingenious (ball/plane, roll/cylinder) to minimize the number of parameters to be studied. The conditions of tests and environment are well controlled with in situ microscopic observations (in a scanning electron microscopy [SEM], for example). These tests are often carried out in research laboratories. Certain fundamental tests are specific tests.
  - **Specific tests:** Specific tests are designed to compare various solutions in conditions close to those of a real tribosystem of certain typical parts, while keeping simplicity for the modes of contact (pin-on-disc, plane/plane, cylinder/plane). Most of the standardized tests are part of specific tests. These tests are often used by the suppliers of materials to evaluate products on the basis of the treatments of surfaces and lubricants on a common basis. They are extensively used, in specifications, by manufacturers to guarantee the tribological features or functionalities of the products coming from the whole world.
  - **Simulation tests**: Simulation tests are tests to validate and verify the features of components. As all the real conditions are simulated they are expected to reproduce the answers of a real tribosystem on a bench test or on a real machine with often whole or partial real parts instead of the standard test samples (piston ring/cylinder liner of engine, gas valve of oxygen tank, joint/car window, and drum brake of aircraft's throttle, for example). The interest of these tests is to reproduce the mechanisms of

wear of the real system by simplifying a certain number of input parameters. The difficulties of these tests reside in the integration of sensors of measurements on real parts without modifying the conditions of mechanical loads.

Tribological tests can be also classified according to their type of measurement: test of friction and test for wear.

The tests of friction are essentially classified according to their mode of contact and the type of movement, while the tests of wear are based on the mode of wear to be produced on the sample.

The usual configurations of tests of friction are:

1/Ball/plane
2/Plan/plane
3/Cylinder/plane
4/Cylinder/cylinder
5/Axis/bearing

These modes can be combined with a movement in rotation or in translation. The tests of wear can be divided into six categories:

- **Abrasion tests:** Abrasive wear is produced by the friction of abrasive materials often harder than the material to be tested. The tracks of wear characterized by macroscopic plastic deformation by forming scratches or chips (Fig. 8.2) (ploughing and cutting phenomena). For the choice of materials for ploughshares, bucket teeth, or shield plates of crushers, for example, we often need to know the abrasion resistance of materials to be used. To compare the abrasion resistance of various materials, tests were designed to investigate wear of the sample in friction against an abrasive material (an abrasive grindstone, for example, abrasion with two bodies) or an antagonist body in an abrasive environment (a steel part or an elastomer in sand, for example, abrasion with three bodies). The important point of these tests is to maintain constant abrasiveness by renewing the abrasive during the test, because the size and angularity modifications of the abrasive during the test can change the aggressiveness of the antagonist.

(a) (b)*

Figure 8.2 (a) Secondary electron image of an abrasive wear and (b) typical abrasive faces of a mechanical part (*http://fr.wikibooks.org/wiki/Tribologie_-_Usure_des_surfaces).

(a) (b)*

Figure 8.3 (a) Secondary electron image of an adhesive wear and (b) typical adhesive surface (*http://fr.wikibooks.org/wiki/Tribologie_-_Usure_des_surfaces).

- **Adhesion tests:** Adhesive wear occurs when a junction formed between two surfaces in contact is more resistant than one of both materials. In that case, the separation of both surfaces in contact takes place in one of the first bodies, which forms a transfer on its antagonist (Fig. 8.3). Wear resistance of a couple of materials depends essentially on their tribological compatibility. The tests of adhesion aim at comparing this resistance according to various configurations of contact.

(a) (b)

Figure 8.4 Erosive wear by particles: (a) global view and (b) detailed view (http://fr.wikibooks.org/wiki/Tribologie_-_Usure_des_surfaces).

They are generally used to compare various solutions of materials and surface treatments. The characteristics of wear debris often have significant influences on the test results.

- **Erosion tests by particles:** The common point between erosion and abrasion is that the antagonist of the studied material is an environment constituted by abrasive particles. But contrary to abrasion, which generates a visible macroscopic plastic deformation in the optical microscope, erosion induces wear in a much lower scale, with few scratches but by leaving visible surface waviness to the naked eye (Fig. 8.4). Wear depends rather on the velocity, on the mass of particles, and on the number of impacts, while abrasion depends on the pressure of contact. The tests of erosion are thus often realized by a jet of abrasive on a plate or by a rotating rotor in an abrasive environment.
- **Cavitation tests:** Cavitation is caused by the impact of microjets at the end of an asymmetrical implosion of bubbles when they go into a zone of overpressure near a solid surface. The tracks of cavitation are characterized by impact craters. Their intensity depends on the variation of the pressure (Fig. 8.5).

Figure 8.5 Cavitation wear (http://fr.wikibooks.org/wiki/Tribologie_-_Usure_des_surfaces).

Figure 8.6 Fretting wear: the presence of red corrosion products (http://fr.wikibooks.org/wiki/Tribologie_-_ Usure_des_surfaces).

The implosion of bubbles is generated either by a vibrating part in a liquid or by a flow of turbulent fluid that causes zones of low pressure and overpressure.

- **Fretting tests:** Fretting is wear by friction due to low-amplitude movement between two mating materials—generally in a stroke from some micrometers to some hundreds of micrometers. Very fine red, ochre, or black wear debris around the track of contact is often observed due to fretting wear (Fig. 8.6). The tests of fretting are conducted either with a ball against a specimen set in a tensile test machine with oscillating load or by a sample submitted to a reciprocal movement by a mechanical system or a shaker.

- **Superficial fatigue tests:** Superficial fatigue is characterized by cracks generated by cyclic stresses of contact. Parallel or perpendicular cracks in the contact area according to the directions of the main stress are noticeable. The tests of superficial fatigue are designed to have a field of required stresses by combining the configuration of contact of both parts, the rate of sliding, and the coefficient of friction by putting some lubricant.

Both types of classifications are complementary according to the objectives of the tests. Sometimes correlations are found between these two types of classifications. For the tests of simulation, for example, it is important to use parts closest possible to real parts under test conditions closest to those of functioning. It is also crucial to reproduce the same mechanisms of wear by comparing their wear faces after the test.

8.2.3 *State of the Art of the Tribometers*

Even if archaeological research shows that the movement of pharaohs' statues on rollers constitutes a valuation method of friction in antiquity, the first measures of friction were realized by Leonardo da Vinci (Fig. 8.7, [3]) during the Renaissance. These principles of measure are still current for some standardized tests (ASTM D1894, D3248, for example).

One of the first tribometers designed by Gustav Hirn, in 1854, consists of measuring the strength of friction by means of a system of balance. This tribometer allowed Hirn to study the relation between the coefficient of friction and the input parameters such as pressure, speed, and contact area in dry and greased contact (Fig. 8.8, [3]). This tribometer already contained the essential

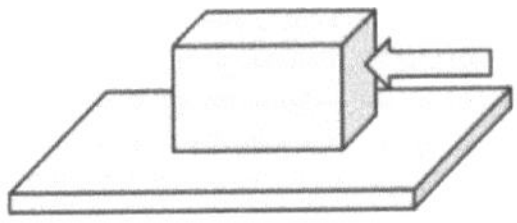

Figure 8.7 Schematic illustration of a measurement of friction on a horizontal plate [3].

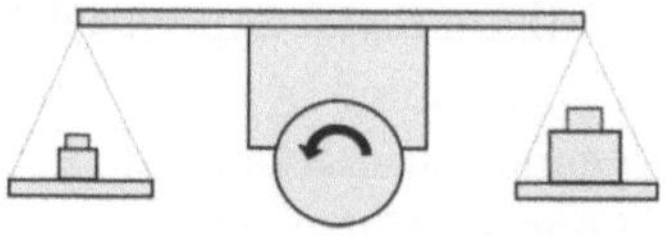

Figure 8.8 Schematic illustration of the Hirn tribometer [3].

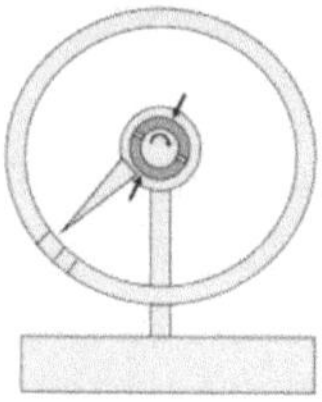

Figure 8.9 Schematic illustration of the Thurson tribometer.

elements: the mechanical drive for the movement, the applied load, and the measurement system by weight.

Robert Thurson developed, in 1885, a tribometer that measures directly the resistive torque of a bogie bearing of a train with the same principle. The efficiency of the measure is improved via a system using an arm lever, which allows one to have a direct reading of the results on a dial (Fig. 8.9).

This type of measurement system of friction was gradually given up with the arrival of various sensors for strength, speed, temperature, and humidity.

Considering the number of parameters to be controlled and the complexity of measures, numerous tribometers were designed and manufactured in the industrial and university laboratories for a given study. A recent inventory realized by the National Physical Laboratory of the United Kingdom counts more than 400 tribological test types in the world [4]. A famous sentence says "There was during a certain number of years so much of tribometers that of tribologists." It was before specialized manufacturers occupied more and more place in the market. These manufacturers now either propose products on catalogs often adapted to the standards or develop tribometers in association with researchers, while taking advantage of their skills acquired on the standard products in the same time.

Today, a typical tribometer consists of four essential elements, which represent a complete tribological system. It is a question of:

- the motorization and the transmission. They put in movement the mobile part with regard to the fixed part with a kind of given movement;
- the loading system. It applies the load to reach required contact pressures at the level of the contact by mass, spring, hydraulic, or pneumatic cylinder;
- the system of measure and acquisition. It measures and records the outputs of the tribosystem (force of friction, wear, vibrations, heating); and
- accessories for the environment of the test. They maintain or regulate the parameters as temperature, humidity, pressure of the gas, flow of lubricant, etc.

The measurement system is the essential element of a tribometer because its reliability depends strongly on the quality of the answers of the system. Here are the essential measures of a tribometer:

- **Friction measurement:** The friction force measured on the sample holder is a resultant of the force of contact generated in the interface that varies versus time and space. Collisions between asperities, cracks by cuttings of the first and third bodies, and impacts of particles in the interface are transmitted to the force sensor via an elastic body. This force is then transformed in an electric signal by the used sensor. The reliability of this measure thus depends on the type of used sensor and on the method of acquisition. Piezoelectric sensors are more sensitive to the fast variations of the friction (unstable friction, vibrations, and squeal noise). The strength generated by an alternate movement allows elimination of the drift. But these drift are predictable with difficulty for unidirectional strengths. The sensors with capacities are more stable at the level of the zero point for this type of measure with a bandwidth often lower than that of a piezoelectric sensor.
- **Wear measurement:** Wear can be estimated by measures:
  - Dimensional by means of thickness sensors or by distance

- Weight by means of a balance
- Mass via radioactive tracers

The weight measures are precise and simple to apply. But they require dismantling of parts, which is difficult for continuous measurement during a long-term test. The dimensional measures do not need a disassembly of the samples. Besides, they give wear evolution during the tests, which is often interesting for nonlinear behavior. But they can be perturbed by thermal dilations due to the heating by friction. The measures by radioactive tracers are independent from thermal disturbances with very precise results of wear for parts with complex geometrical shapes on which the dimensional measures are applicable with difficulty. The techniques of measure and the equipment in that case are heavier to manage than the dimensional techniques.

- **Noise and vibration measurement:** Vibrations and noises generated by friction instability are a problem that worries numerous tribologists, not only by their consequences on the environment (noise during the braking for example), but also by the disturbances on the measurements. The generation of these instabilities depends at the same time on variations of the friction force versus time and on the rigidity of the system (contact stiffness).
  The measurement of vibrations by an accelerometer allows having a first idea on friction instability before making measures of deformations and air noise.
- **Temperature measurement:** The measure of temperature of parts in friction is often difficult because of strong gradient of temperature, in particular for insulating materials such as ceramic and polymers.
  The influence of temperature on the tribological behavior of polymers is particularly important because their glass transition temperature (T_g) is relatively low. Two techniques (thermocouple and infrared camera) are usually used. The measurement by a thermocouple is sometimes difficult for ceramic and polymers because of the low thermal conductivity of these materials. The measures by infrared cameras give a global vision of the gradient of temperature

near the zone of friction. But the absolute values depend on thermal emissivity, which varies with temperature.

- **Humidity measurement:** Humidity is very variable according to the weather and geographical conditions. The tribological behavior of certain materials such as PA66 is particularly sensitive to the variation of humidity. The behavior of certain metallic pairs can also change with humidity because of the modifications of the rheological behavior of the third body.
- **Electrical measurement:** The measure of the electric resistance of the contact with an imposed voltage allows having useful information on the thickness of an oil film, the formation of oxides, and wear of an insulating deposit. Certain standards in the automotive industry require this measure for the electric contacts.
  The measure of static electricity for insulating pieces is interesting in order to know their role in the friction, which can be also be important as that of mechanical origin. Numerous studies were conducted in the domains of packaging and package distribution.

Annexure 8.1 gives the standard test in tribology.

8.2.4 *Methods of Presentation of Tribological Results*

The results of tribological tests are the values of friction and wear, which are useful data because they directly arise from test conditions or conditions of the tribosystem. They can be presented by diverse manners. Classically, these tribological or output data can be shown in the form of frictiograms, diagrams of transition, or maps of wear.

Frictiograms (even "tribographs") are graphic representations of a measured quantity of friction or wear versus operating parameters (e.g., normal load F_N, speed v, temperature T, duration of the test t and the number of cycles or the sliding distance); structural parameters (e.g., the couples of tested materials, hardness, roughness, or even microstructure); and parameters of interaction (e.g., contact stress, thicknesses ratio-roughness, or modes of lubrication).

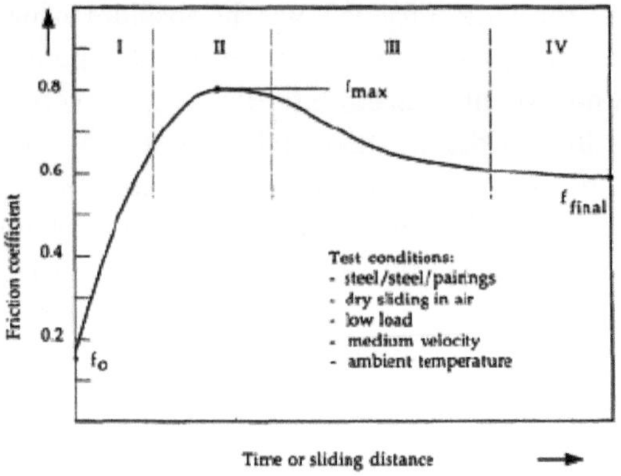

Figure 8.10 Example of a frictiogram for a metallic pair in drysliding conditions [5].

The diagrams of transition allow identifying the critical values of the operating parameters for which a regime of wear and friction turns out to be catastrophic.

The maps of wear are the third mode of presentation of the data, which characterize the various regimes of damage of surfaces.

The presentation of the friction and wear data is illustrated here with some typical examples.

8.2.4.1 Temporal dependence of tribological data

The simplest test of tribology always gives the following information: a set of operating parameters (input parameters) defined and constant and the values of friction and wear (output parameters) versus time. Figure 8.10 shows the typical curve between the coefficient of friction, μ, versus time ($\mu = f(t)$) during dry sliding involving a couple of metallic materials [5]. The curve can be cut in four successive stages:

(1) At stage 1, the initial value of the coefficient of friction, which is usually 0.1, is dependent on low loads F_N and on shear resistance of the surface contaminants. On the other hand, it is independent from the grades of the couple of tested materials. The removal of the superficial layers results in the increase of adhesion further to the appearance of interfacial areas'

exposure. In addition, growing interactions of asperities of both materials in contact as well as the rising production of wear debris that can be trapped lead to a gradual increase of the coefficient of friction. Stage 1 is a running-in period for the surfaces.

(2) At stage 2, the maximal value of the coefficient of friction (μ_{max}) from 0.3 to 1.0 for the most part of the metallic pairs is reached with the maximal interfacial adhesion, the deformation of the asperities, and the more important trapping of the wear particles.

(3) At stage 3, a decrease of the coefficient of friction can appear further to the potential formation of a protective tribochemical layer as well as a lesser damage of the asperities of contact.

(4) Stage 4 is characterized by a quasi-stationary state leading possibly to constant values of coefficients of friction.
The frictiogram (Fig. 8.10) shows a smoothed graph, which, in practice, can be much more fluctuating with peaks of friction, variations due to effects of sticks-lip type, for example, or still to vibrations of the test machine, which must be verified on this point systematically (Fig. 8.11). Besides the presentation of the frictiogram, friction behavior can be characterized by average data:

- of the coefficient of friction μ;
- of the coefficient of friction μ_{max};
- of the coefficient of friction at the end of test μ_{final};
- of the coefficient of average friction; and
- of the standard deviation on the coefficient of average friction.

As a general rule, the tests of friction and wear are repeated at least twice by modalities for the reproducibility of these values. Figure 8.12 gives in, example another shape of the comparative graph, taking into account structural parameters (couples of different materials) [5].

Concerning the curves of wear versus time, a classic curve is shown in Fig. 8.13 always for a couple of metallic materials in dry sliding [5].

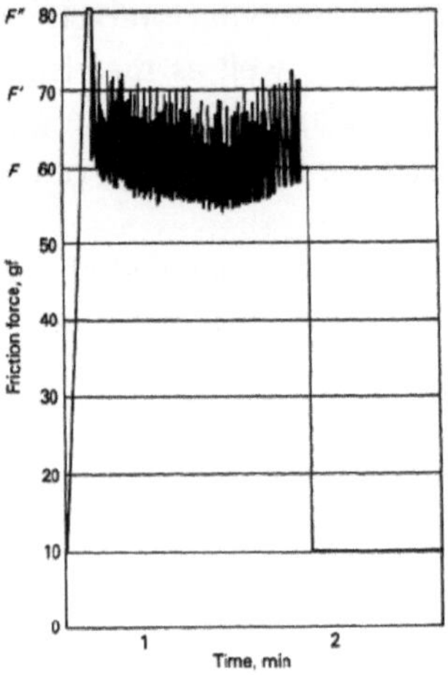

Figure 8.11 Example of a no-smoothed frictiogram: stick-slip behavior [5].

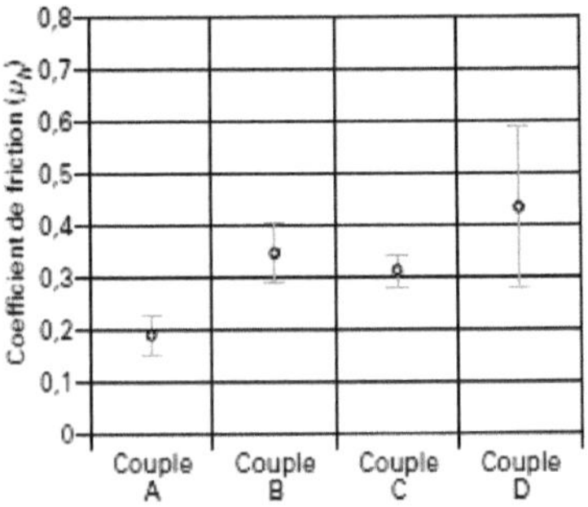

Figure 8.12 Comparative graph of friction coefficient versus pair of materials [5].

In the same way as shown previously, the curve can be divided into three stages:

In stage 1, an incubation time where no damage of wear appears can precede it. In the simplest cases, a wear law can be used:

$$\frac{dW}{dt} \approx \frac{1}{W} \rightarrow W(t) = \text{const } t^{1/2} \tag{8.1}$$

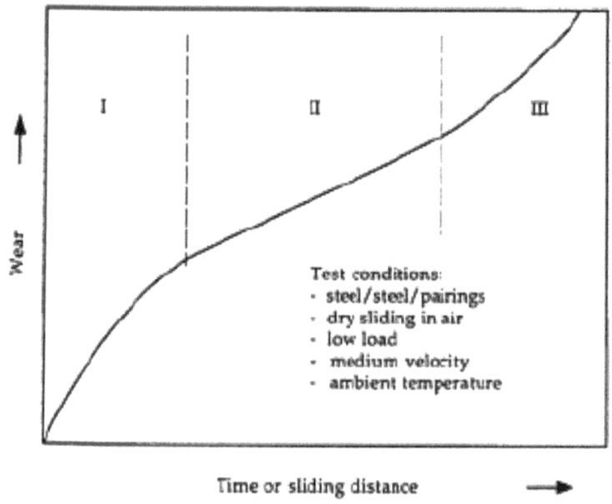

Figure 8.13 Example of wear curve [5].

In stage 2, the tribosystem can also show an almost linear behavior. The law of wear is then specified as

$$\frac{dW}{dt} = \text{const} \rightarrow W(t) = \text{const}\ t \tag{8.2}$$

For technological applications, the bearable values of the wear rate k must be lower than 10^{-6} mm^3/N.m or (MPa^{-1}).

In stage 3, an acceleration of wear can occur with a pooling of a certain number of wear mechanisms until failure of the contact. The wear law, in a simplified way, can then take the following shape:

$$\frac{dW}{dt} \approx W \rightarrow W(t) \approx e^{\text{const}\ t} \tag{8.3}$$

8.2.4.2 Dependence of tribological data compared to input parameters

Besides temporal representations, tribological data can be drawn versus the parameters of the tribosysteme such as normal load F_N, sliding speed v, movement or the others, temperature T, hardness, etc.

Figure 8.14 shows another way of representing tribological data via a polar diagram [3, 6], in particular when surface treatment is investigated. It is enough to draw a reference circle of radius 1 and to divide it into four sectors. A number of relevant parameters are then defined. Every parameter is a ratio between the treated surface and the reference surface.

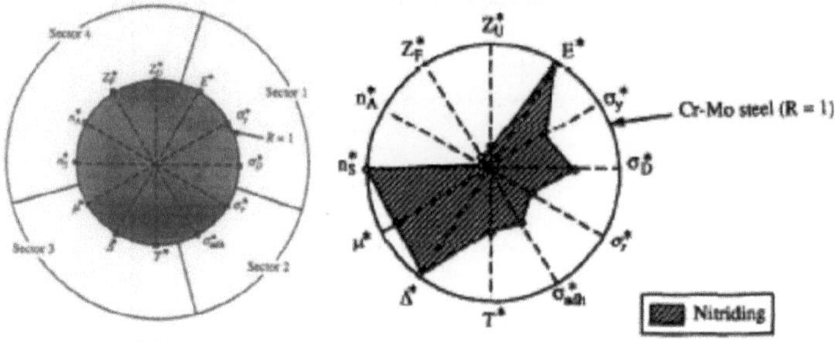

Figure 8.14 (a) Principle of a polar diagram and (b) example of surface treatment by nitriding [6].

The first sector concerned the intrinsic properties such as Young's modulus, type of surface treatment of the surface zone involved (thin or thick), yield limit, and endurance limit. The second sector indicates the effects of the interaction between the treated zone and the substrate. Residual stresses and adhesion stresses can be also considered.

The third sector is focused on the local solicitations with, for example, the rate of load and the coefficient of friction. The fourth sector gives the output parameters of the tribological system, associated parameters in the answer of the material as the number of cycles to the initiation of cracks, the worn depth, etc.

This type of graph allows visualizing in an efficient way, compared to a reference couple and a given tribosystem, and the performances of solutions' antifriction and antiwear. At the same time as these representations and to estimate the trends of the wear and friction results obtained, the application of data processing by data mining allows this type of crossed presentation (Fig. 8.15, [7]).

8.2.4.3 Wear map

Another way to understand wear mechanisms is by developing wear maps, resuming the results of experimental tribological studies [8, 9]. Wear maps are in two (2D) or three dimensions (3D) with also dimensionless parameters.

They synthetically allow visualizing of all the results according to the most important tribological parameters. They can be even retained as a reference table to predict a tribological behavior (with all the usual precautions). Figure 8.16 gives an example of 2D and 3D representations of wear maps of a diverse case [5, 7].

8.2.4.4 Transition diagram

To identify efficient or critical regimes of tribosystems, diagrams of transition can be utilized. For example, for greased Hertzian metallic contacts, a methodology in three stages for the construction of such a diagram was developed by the International Research Group one Wear of Engineering Materials (IRG-OECD) [5].

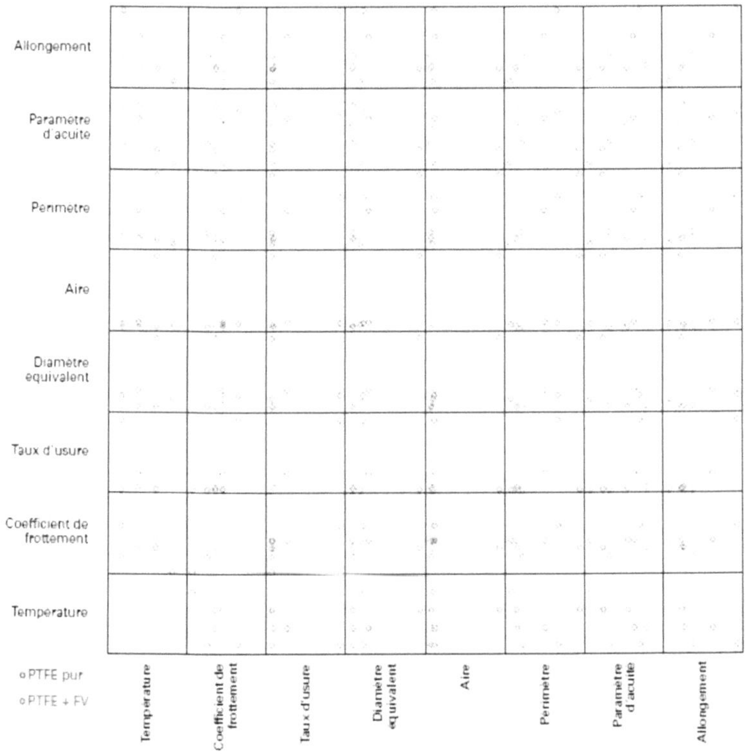

Figure 8.15 Data mining of tribological parameters [7].

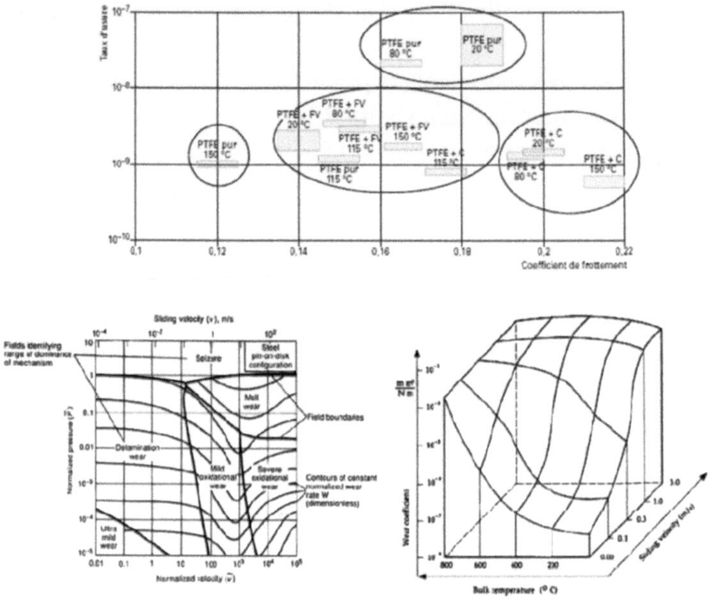

Figure 8.16 Examples of 2D and 3D wear maps [5, 7].

Stage 1: All the tests are perfectly defined (complete definition of the tribosystem). The tests can be carried out either with new samples at each step of loading or with a single couple of materials. In every experiment, the coefficient of friction and the wear rate are assessed.

Stage 2: For a set of given parameters (speed, temperature, pressure, time) the critical load F_{Ncrit} or threshold between two regimes is characterized, as in Fig. 817a.

Stage 3: The critical load F_{Ncrit} for a variation of sliding speed v, the other input parameters being constant, is determined. The graph of the normal critical load F_{Ncrit} versus sliding speeds v can be also named "diagram of transition IRG" (Fig. 8.17b).

To complete the tribological test results obtained somehow, these are traditionally associated to microscopic, physicochemical analyses and also to fine, appropriate spectrometries to surfaces (X-ray photoelectron spectrometry (XPS) and glow discharge mass spectrometry (GDMS)). All the results can be then collected in databases [10].

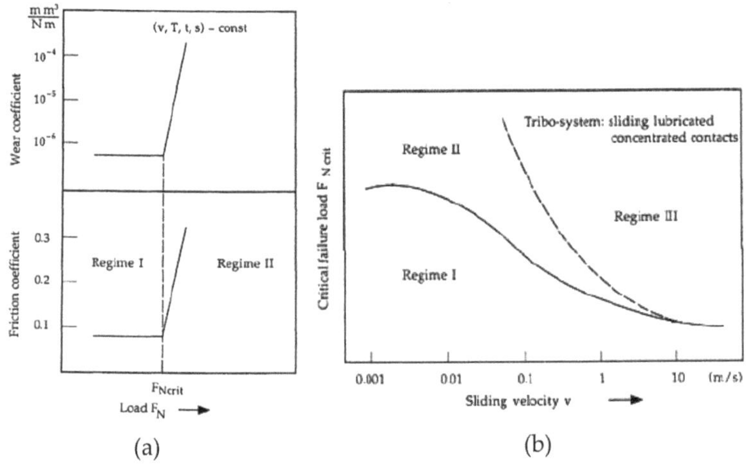

Figure 8.17 Assessment of the critical load for lubricated contacts under dry-sliding conditions (a) and example of an IRG transition diagram (b) [5].

8.3 Specific Tribological Measurements: Example of Biomaterials (with and without Surface Treatments)

Most of the artificial joints made of metals and polymers used in the human body fail due to wear and accelerated corrosion [11, 12]. The problems associated with prosthetic failure, such as osteolysis, implant loosening, and consequently revision surgery, is one of the major clinical issues that is highly challenging due to an increase in the aged population and hence total joint replacements (TJRs) [11]. Three types of wear, namely abrasive wear of balls, adhesive and fatigue wear of polymer cups often encountered with artificial hip joints, and abrasive and fretting wear in the case of knee implants, are encoutered [13]. New material developed to be used for artificial joints is tested for the wear resistance and friction coefficient in appropriate simulated body conditions.

Two types of tests are often performed as per ASTM/ISO standards and the knowledge obtained through these laboratory tests is important in the validation of prostheses. Initially during the development of new materials screening tests are performed to select a few materials, that perform well in harsh conditions.

However, it is important to note that screening tests performed as specified by ASTM standards do not lead to results corresponding to real systems as they do not represent implant geometry that actually dictates the contact stress and the type of lubrication. As a result, these tests do not reproduce the wear rate that occurs between pairs of materials used in vivo. The most commonly used screening tests are pin on disc, pin on flat, ring on flat, and crossed cylinders [14] (Fig. 8.18). These machines are often modified to some extent on the basis of the applications, and the wear rate and friction obtained for various titanium alloys obtained using different test conditions are well elaborated by Long et al. [12]. As the pin on disc does not correspond to the reciprocating motion of the joints, the pin on flat reciprocatory wear testing is commonly used to evaluate the wear resistance and friction coefficient of the both coated and uncoated substrates [11, 12].

The second type of test is carried out using a simulator that is designed in such a way that it causes a prosthesis to undergo wear in a similar manner as will take place when implanted in the human body. To accomplish this the simulator provides controlled conditions such as a set of motions, loads, and lubricants almost equivalent to in vivo conditions. The simulator is designed as specified by ISO standards 14242 and 14243. The various types of simulators available in the market worldwide have been reported by Affatato et al. [14] in their review article (Fig. 8.18). Human joints are complex bearings that effectively operate under both fluid film and boundary lubrications normally, and during walking the presence of mixed lubrication, that is, coexistence of fluid and boundary lubrication, is observed [15]. In the case of human joints, the lubrication is provided by synovial fluid, which is a dialysate of blood plasma that is devoid of clotting factors, erythrocytes, and hemoglobin and contains proteoglycans, glycoproteins, and phospholipids, while bovine serum or hyaluronic acid serves as a lubricant in hip simulators and the wear rate is measured using the weight loss method. However, there are slight discrepancies between the wear rate obtained through joint simulators and explanted implants. This is attributed to the method of accessing the wear rate in the case of the failed implant. The usual practice to determine the wear rate is gravimetric, geometrical, volumetric, or image processing [16]. Further, there are several factors that affect

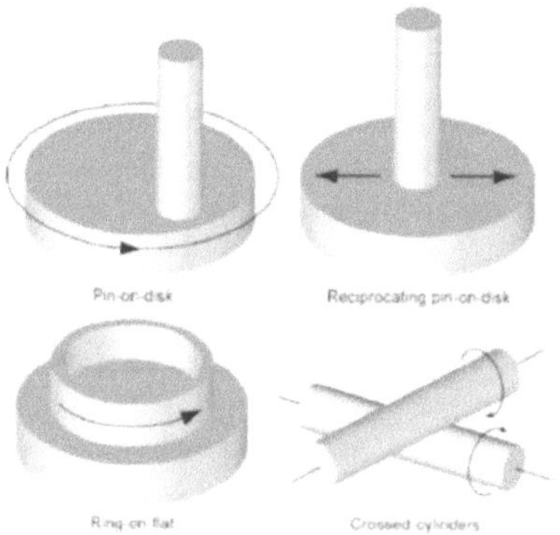

(a) Soft often used wear screening tests

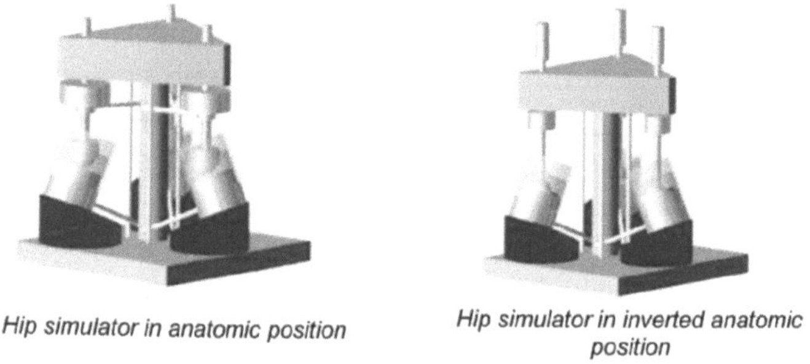

(b) These machines simulate physiological conditions and represent complex dynamic situations

Figure 8.18 Screening test for testing biomaterials and bench tests for the hip joint [14].

in vivo wear with respect to more controlled laboratory tests so that a correlation between in vivo and in vitro data still remains a major challenge. Recently many metal-on-metal implants were observed to shed excess wear debris and failed due to several reasons [17] in spite of the fact that these materials well performed in hip simulator tests. Thus it is very important to modify and develop new simulators that will replicate in vivo behavior as much as possible in order to avoid failures.

The effect of corrosion on wear was not taken into consideration in the past and normally both wear and corrosion testing is performed independently and corrosion and wear rates are reported disjointedly. However, in the recent past more emphasis on measurement of tribocorrosion of materials has been practiced so that the synergic effect of both phenomena is taken into consideration while evaluating the behavior of an implant. Attempts are made to develop an instrument called the tribocorrosion tester, which can measure corrosion and wear simultaneously [18]. Recently, an integrated tribocorrosion cell was developed both in the pin-on-flat testing machine and also in the simulator by Yu et al. in order to evaluate the tribological and electrochemical behavior of the implant material [19]. Their studies clearly demonstrated that 27% of loss of material in the presence of lubricant occurred due to corrosion-related damage.

In addition to the above, efforts are made to understand the effect of roughness on wear behavior of the implant materials by performing nanotribological studies. Nanoscale single-asperity contact and surface topography are examined using atomic force microscopy to understand the effect of roughness and debris on wear mechanisms [20–23].

Currently coatings of the implant are considered as an alternate solution to improve wear resistance of metallic implants. Coatings of TiN, CrN on a CoCr head, and titanium alloys are been clinically proved to enhance the hardness and hence the wear resistance. However, failure of the thin coating clinically has resulted in a look for other ceramic coatings that can have higher adhesion and a longer life span. Diamond and diamond-like coatings (DLCs) are also considered as they provide high hardness and low friction [24]. Recently the wear behavior of nanoceramic alumina + titania and zirconia-coated Ti alloys investigated using the pin-on-plate test showed very promising results [22, 23]. The bilayered coating consisting of zirconia as the first layer and a mixture of alumina + 13 wt% titania exhibited higher wear resistance [22, 23] amongst various types of coatings.

However, one has to be very careful in recommending a coating for biomedical applications, as it was clearly evident with DLC coatings that showed higher wear resistance in simulator tests failed in *in vivo*. The testing protocols are to be modified in such a manner that the testing conditions take into consideration all the variables that are encountered in the human body.

Annexure 8.1: Standard Tests (Often Used)

Configuration	Test objectives	Range of application	Standard
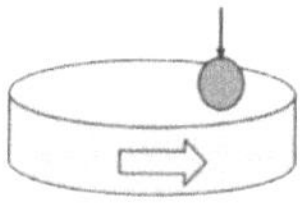	Measurement of friction coefficient and wear in continuous motion	Ball bearings, ball bearing slides	ASM G-99, DIN 50326, ISO 7148
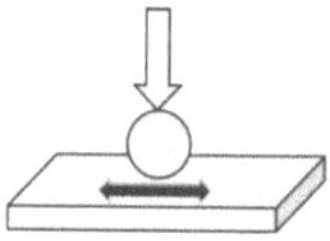	Measurement of friction coefficient and wear in reciprocating motion	Ball bearings, ball bearing slides, autolubricant materials, lubricants, liners/rings	ASM G-133, ISO 7148, G-181
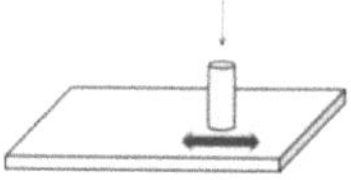	Wear measurement by abrasion	Work tools for soils, mining, quarrying, public works, agricultural machinery	ASTM G-132
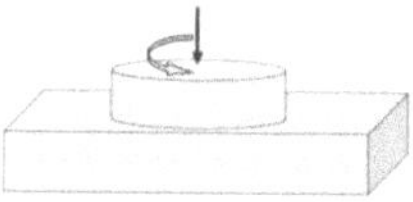	Seizure threshold measurement for a couple of materials	Valves, pumps, slides	ASTM G-98
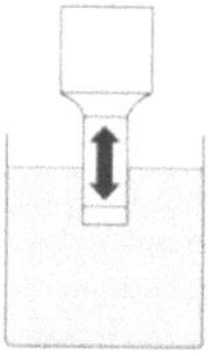	Cavitation test by ultrasonic method	Bearings, valves, pumps	ASTM G-32
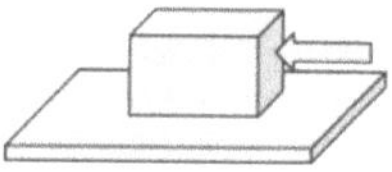	Friction coefficient test in surface configuration	Polymer films, wood, particle boards, pavement	ASTM D1894, D2394, D3247, NF P 98-220-1, C1028
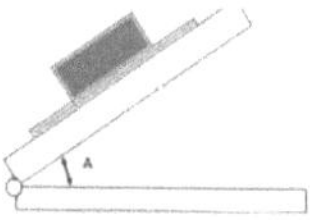	Static friction coefficient test (inclined plane method)	Fiberboard, coatings	ASTM D3248, D3334, D4518
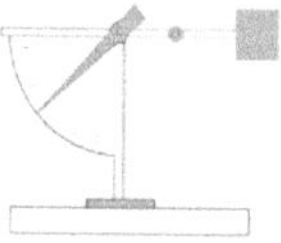	Surface frictional property measurement using the British pendulum tester	Wood, rubber, pavement	ASTM D4103, EN 13036-4

Configuration	Test objectives	Range of application	Standard
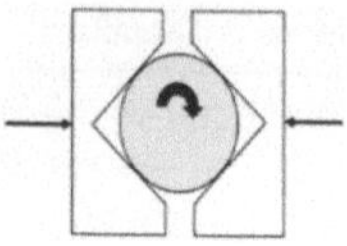	Four balls test	Lubricants, greases	ASTM D2266, D4172, D5183, D2783
	Pin and Vee block test	Lubricants, greases	ASTM D2670
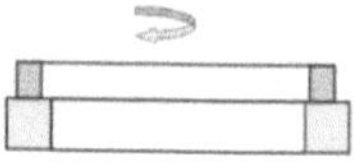	Ring-on-ring test	Autolubricant materials for mechanical stop	ASTM D3702
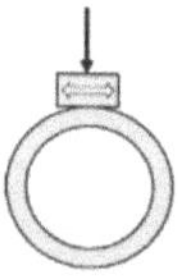	Block-on ring test	Materials, lubricants, greases	ASTM D3704, G-77, G-137
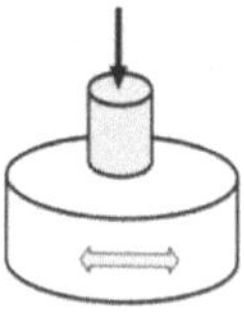	Pin-on-flat test	Polymer materials for prostheses	ASTM F732
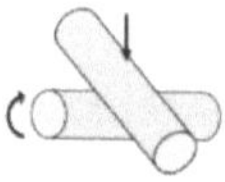	Crossed-cylinder test	Metallic materials	ASTM G-83
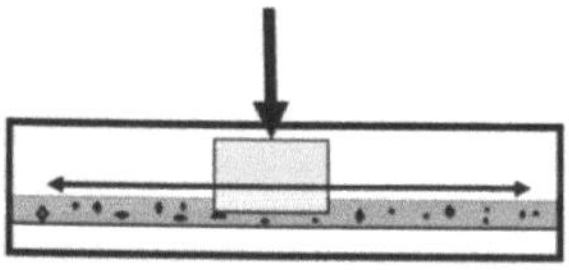	Abrasion test, Miller test	Aggregates abrasiveness, abrasion resistance of materials	ASTM G-75
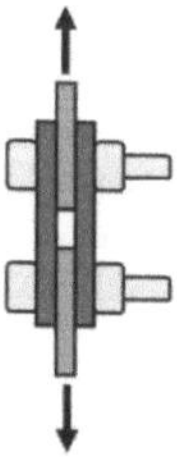	Friction coefficient measurement for bolted parts		NF P22-461

Configuration	Test objectives	Range of application	Standard
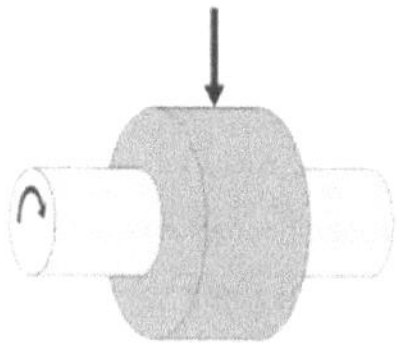	Tribological behavior for journal bearings	Antifriction materials	ISO 7148-2
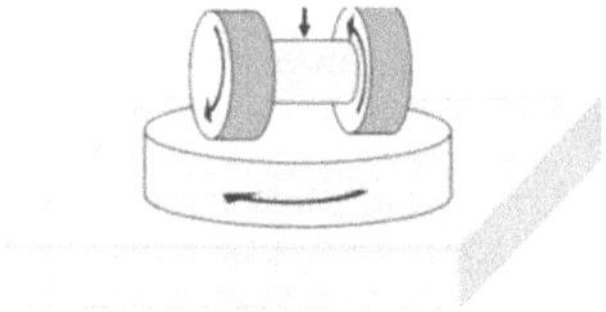	Abrasion Taber test	Materials, coatings	ASTM F-1978
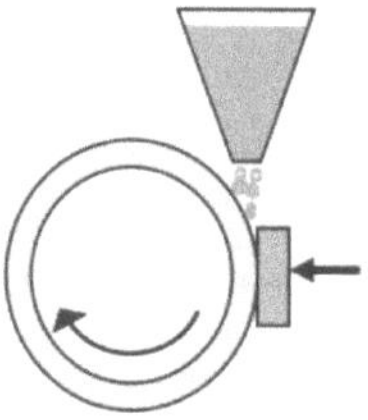	Abrasion test	Materials, coatings	ASTM G-65

References

1. Belot, J.-M., and Rigaut, B. (1994). Importance economique de l'usure, *Cetim Inf.,* **141** (in French).
2. Czichos, H (1978). *Tribology* (Elsevier).
3. Dowson, D. (1998). *History of Tribology* (Professional Engineering).
4. Neal, M.J. and Gee, M. (2000). *Guide to Wear Problems and Testing for Industry* (William Andrew).
5. Czichos, H. (1997). *Presentation of Friction and Wear Data, Friction and Wear testing,* source book of selected references from ASTM standards and ASM handbooks, 47–50.
6. Carton, J.F, Vannes, A-B., and Vincent, L. (1995). Basis of a coating choice methodology in fretting, *Wear,* **185**, 45–57.
7. Kowandy, C., Richard, C., and Chen, Y.M. (2008). Characterization of wear particlesfor comprehension of wear mechanisms: case of against cast iron *Wear*, **265,** 174– 179.

8. Chen, Y.-M., Rigaut, B., Pavy, J.-Cl., and Armanet, F. (2002). Wear particles forming byphase transformation in PSZ ceramics during-high speed sliding in, *Proceedings of the Leeds-Lyon Symposium* (Elsevier).
9. Chen, Y.-M., Rigaut, B., and Pavy, J.-Cl. (1992). Usure des céramiques à grandes vitesses, *Cetim Inf.*, **126** (in French).
10. Woydt, M. (2004). Application des données tribologiques des matériaux *Tech l'Ing*, Réf. BM7006 (in French).
11. Geetha, M., Singh, A.K., Asokamani, R., and Gogia, A.K. (2009). Ti based biomaterials, the ultimate choice for orthopaedic implants: a review *Prog. Mater. Sci.*, **54**, 397–425.
12. Long, M., and Rack, H.J. (1998). Titanium alloys in total joint replacement materials science perspective, *Biomaterials*, **19**, 1621–1639.
13. Dowson, D. (1995). A comparative study of the performance of metallic and ceramic femoral head components in total replacement hip joints, *Wear,* **190**, 171–183.
14. Affatato, S., Spinelli, M., Zavalloni, M., Mazzega-Fabbro, C., and Viceconti, M. (2008). Tribology and total hip replacement: current concepts in mechanical simulation, *Med. Eng. Phys.* **30**, 1305–1317.
15. Neu, C.P., Komwopoulus, K., and Reddi, H. (2008). The interface of functional biotribology and regenerative medicine in synovial joints, *Tissue Eng.*, **14**, 235–247.
16. Dowson, D. (2001). New joints for the millennium: wear control in total replacement hip joints, *Proc. Inst. Mech. Eng. H*, **215**, 335–358.
17. Smith, A.J., Dieppe, P., Vernon, K., Porter, M., and Blom, A.W. (2012). Failure rates of stemmed metal-on-metal hip replacements: analysis of data from the National Joint Registry of England and Wales, *Lancet*, **379**, 1199–1204.
18. Diomidis, N., Mischler, S., More, N.S., and Roy, M. (2012). Tribo-electrochemical characterization of metallic biomaterials for total joint replacement, *Acta Biomater*, **8**, 852–859.
19. Yana, Y., Dowson, D., and Neville, A. (2013). In-situ electrochemical study of interaction of tribology and corrosion in artificial hip prosthesis simulators, *J. Mech. Behav. Biomed. Mater.*, **18**, 191–199.
20. Ho, S-P., Carpick, R.W., Boland, T., and LaBerge, M. (2002). Nanotribology of CoCr-UHMWPE TJR prosthesis using atomic force microscopy, *Wear*, **253**, 1145–1155.
21. Check, J., Karuppiah, K.S., and Sundararajan, S. (2005). Comparison of the effect of surface roughness on micro/nanotribological behavior of

ultra-high-molecular weight polyethylene (UHMWPE) in air and bovine serum solution, *J. Biomed. Mater. Res. A*, **74**(4), 687–695.

22. Sathish, S., Geetha, M., Aruna, S.T., Balaji, N., Rajam, K.S., and Asokamani, R. (2011). Sliding wear behavior of plasma sprayed nanoceramic coatings for biomedical applications, *Wear*, **271**, 934–941.
23. Sathish, S., Geetha, M., Aruna, S.T., Balaji, N., Rajam, K.S., and Asokamani, R. (2011). Studies on plasma sprayed bi-layered ceramic coating on bio-medical Ti–13Nb–13Zr alloy, *Ceram. Int.*, **37**, 1333–1339.
24. Holmberg, K., and Mathews, A. (2009). Coatings tribology: mechanisms, techniques, properties and applications in surface engineering, *Tribol. Interface Eng.*, Ser. 56, **10**, 433–435.

Chapter 9

Elastic Properties Characterization by Means of Dynamic Resonant Technique

Pascal Gadaud

Institut P', ISAE-ENSMA, DPMM, 1 avenue Clément Ader, Téléport 2, BP 10109, F86961 Futuroscope Cedex, France

pascal.gadaud@ensma.fr

The determination of elastic properties by means of a nondestructive resonant vibrating method has been developed for bulk and coated materials characterization. The study of bulk materials is investigated in relationship with physical properties. Analytical models of composites vibrations allow us to determine accurately the modulus of both the coating and the substrate and applied to single-coated materials; this formalism is then extended to the study of multicoated materials. For materials presenting gradients of mechanical properties, a simplified n-layer model allows to estimate the modulus gradient. Recent developments concern the determination of the shear modulus and the evaluation of residual stresses for bulk and coated materials.

Materials Characterization: Modern Methods and Applications
Edited by Narayanaswami Ranganathan

ISBN 978-981-4613-06-4 (Hardcover), 978-981-4613-07-1 (eBook)
www.panstanford.com

9.1 Introduction

The use of dynamical resonant methods to determine the constants of elasticity of bulk materials has been formalized 40 years ago by Spinner et al. [1–3]. These authors determined accurately all the analytic formulas, correcting factors, and experimental features to supply for flexural, longitudinal, and torsional modes on different specimen shapes; it gave rise to an ASTM standard test method updated in 2001 [4]. It is a convenient alternative between classical quasistatic and ultrasonic tests, which has brought a lot of results in materials science for standard materials characterization. Different excitation and detection devices have been then developed to cover the largest range of materials under various temperatures [5, 6]; recently, progresses in signal processing and in network analyzer technology allow the measurements to be more accurate than ever.

Furthermore, the development of new materials (superalloys, polymers, ceramics, composites) and structural coatings requires new needs in nondestructive techniques to characterize the elasticity in relation to engineering applications: high-temperature use, structure optimization, or shaping processes. The aim of this chapter is to point out the experimental and theoretical developments made on the dynamical resonant technique to obtain a relevant tool to answer these needs.

9.2 Elasticty of Bulk Materials

9.2.1 *Formalism-Experimental Features*

The Young's modulus determination is based on the bending vibration of a beam specimen under frequency continuous excitation [7]. It allows us to perform experiments at 1 K/mn under high vacuum (10^{-4} Pa) from 150 up to 1400 K without any harmful contact, the sample being maintained horizontally between steel wires located at the vibration nodes. Furthermore, excitation and detection are ensured by an electrostatic device (capacitance created between the sample and a unique electrode). The Young's modulus E of a bulk sample in bending mode is directly determined by the relation (longitudinal modulus):

$$E = 0.9464\rho F^2 \frac{L^4}{h^2} T\left({}^{h}/_{L}\ \nu\right) \quad (9.1)$$

where F is the resonance frequency, ρ the density, ν the Poisson's ratio, h and L the beam thickness (1 to 2 mm) and span length (20 to 30 mm), and $T(h/L, \nu)$ a correcting factor close to 1 [2, 3].

Nonconducting samples must be plated on one face in order to be electrostatically excited. If metallic vapor deposition (about 20 nm of AuPd) is usually employed [8], unfortunately AuPd films do not remain adhesive above 600°C. Therefore, Cr sputtering is preferably used for higher temperatures studies but requires a specific setup.

Another testing head allows to determine simultaneously the Young's modulus E and the Poisson's ratio ν on cylindrical specimens (length L ($\sim$50 mm) and diameter d ($\sim$5 mm)) by use of longitudinal harmonic waves [7]. The relation between the resonance frequency F_i of the i harmonic and the two constants of elasticity is

$$F_i = \left[\frac{E}{\rho L^2}\left(1 - \frac{1}{2}\left(\frac{i\pi\nu d}{L}\right)^2\right)\right]^{1/2} \quad (9.2)$$

Practically, two frequencies (the fundamental and the first or second harmonic) are measured allowing to determine ν and then this value injected to determine E.

Starting from the testing head developed previously to determine the Young's modulus in bending mode in the kilohertz range and in agreement with ASTM recommendations, a new testing head has been designed (Fig. 9.1) to determine the shear modulus of plates (30 × 12 × 1.5 mm^3 typical dimensions) in torsion mode [9]. To avoid any harmful contact, the sample beam is maintained horizontally between crossed steel wires located along the vibration node lines.

Torsional deformations of bars with a rectangular cross section have been early studied by Timoshenko [10]. It depends upon a complex shape factor function of the thickness/width ratio resolved numerically [2, 11, 12] and is only applicable to bulk materials. To extend elastic characterization to coated materials, an exact analytic equation of vibration has been developed when the thickness/width

Figure 9.1 Testing head for torsional vibrations.

ratio is small [9].

$$F^2 = \frac{\rho}{G}\frac{1}{L^2}\frac{B^2}{H^2+B^2}\left(1-\frac{B}{\sqrt{3}H}\right) = \frac{\rho}{G}\frac{4}{RL^2} \quad (9.3)$$

where G is the shear modulus and H the width of the specimen. R is the shape factor mentioned above.

For our standard sample dimensions, $B/H = 1/8$ and R takes the value 17.51. This value is in good agreement with literature data [2, 11, 12], giving numerical values from 17.38 to 17.64 for the same dimensions.

9.2.2 *Metals*

The modulus of metals is often studied as a function of temperature up to the range of their use or their shaping; however, some variations of this mechanical characteristic exist with industrial materials because of microstructural variations such as texture or anisotropy, coming from casting or forming.

The case of macroscopic anisotropy has been studied with a quasi-single-crystalline biphased superalloy (coherent precipitates of γ'Ni3Al ordered face-centered cubic [fcc] phase in a matrix of γ

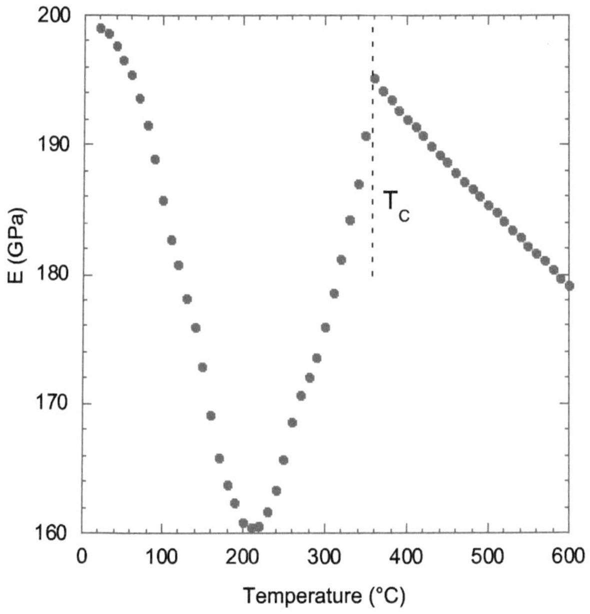

Figure 9.4 Young's modulus evolution of pure Ni as a function of temperature.

Fe, Ni, or Co but also exists for compounds. This effect can be explained by domains theory [14]: the anisotropic magnetostriction in magnetic domains changes drastically the energy equilibrium and so the elastic constants. Figure 9.4 shows this effect on the Young's modulus of pure Ni studied in flexural mode. From room temperature up to 200°C, the ΔE effect increases because the magnetostriction anisotropic factor increases, and then up to T_C, thermal agitation decreases the magnetic alignment and softens the ΔE effect. We must note that this effect is only observable with very low-amplitude load tests as the resonant method or US (applied strain in the range 10^{-7}–10^{-6}), it completely disappears under higher amplitude load (quasistatic tests) saturating local spontaneous magnetization.

This kind of effect is also observable on antiferromagnetic materials below their Néel temperature.

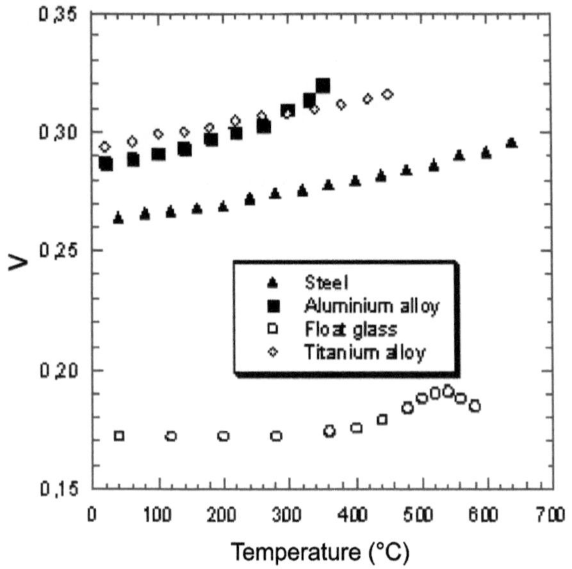

Figure 9.5 Evolution of Poisson's ratio for various isotropic bulk materials.

The last example of measurements on bulk materials concerns the evolution of Poisson's ratio ν with temperature on various isotropic materials. If the standard value of 0.3 is often taken for metals, it is important to determine accurate values and to follow their evolution as a function of temperature.

If longitudinal tests are preferably employed to characterize anisotropic bulk materials (see Section 9.2.2), the more precise method to study isotropic ones is to measure separately E and G, respectively, with flexural and torsional tests and then to deduce ν. Different isotropic materials have been tested (Fig. 9.5) in the widest ranges of rigidity and temperature. As it is impossible to obtain E and G at exactly the same temperature, E/G versus temperature curves are fitted by polynomial laws and ν is calculated at constant steps.

The scattering on Poisson's ratio values determination is in the order of 0.001. The obtained values are coherent with literature data, especially for lower values illustrated by a float glass [15] or SeGe glasses [16].

9.2.3 *Amorphous Materials*

The Young's modulus of different polymers has been determined with flexural tests from vitreous to rubber-like state in the 1 to 10 GPa range. The accuracy of the measurements is better than 0.5% whatever the stiffness. As the resonance frequency is high enough (2–10 kHz) and the imposed stress low enough to avoid dissipative mechanisms [17], the comparison with ultrasonic measurements is satisfying [18] and allows us to validate the method for polymer characterization.

Different studies focussed on the influence of reinforcements on the global elasticity. The interest of these reinforcements is mainly to increase the mechanical properties (polyurethane + glass fibers) but also to lower the elaboration cost (polystyrene + chalk) or to answer ecological considerations (polyurethane + hemp fibers). Figure 9.6 shows the case of a pure semicrystalline polypropylene and reinforced by 30% in volume of wood or straw short fibers ($\leq$500 μm) in view to color itself; the increase of modulus with

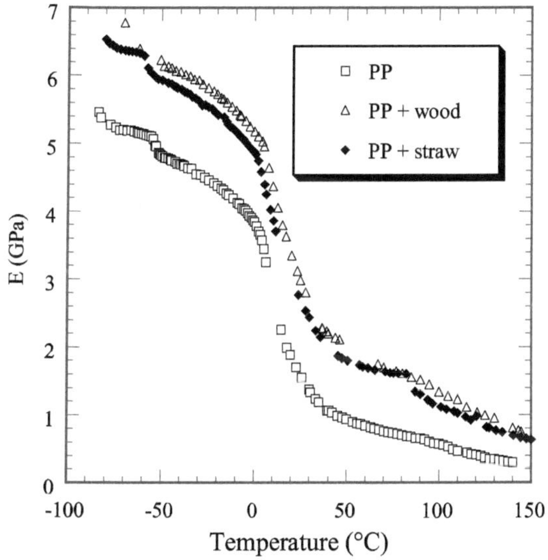

Figure 9.6 Young's modulus versus temperature curves for three grades of polypropylene: pure, reinforced by straw, and reinforced by wood short fibers.

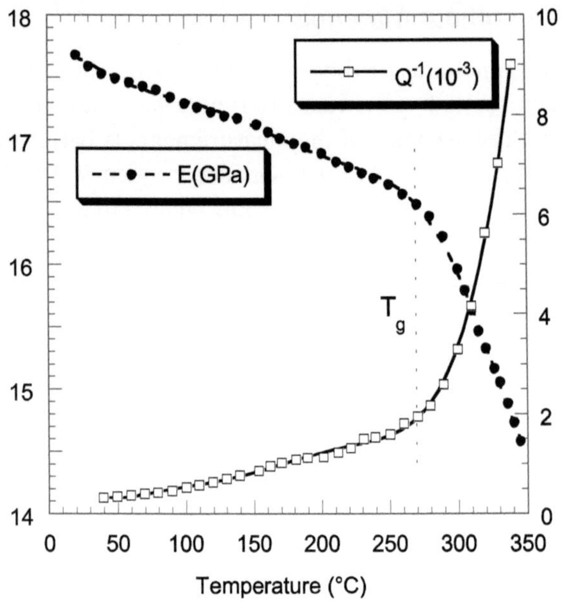

Figure 9.7 Evolution of the modulus and damping coefficient of a GeAsSe glass; T_g is the glass transition temperature.

fibers is, respectively, 56% and 77% at room temperature. The characteristic temperatures (secondary relaxation and α transition) are not changed by addition of fibers, as it is generally assumed.

The general shape of the modulus versus temperature curves of glasses is illustrated in Fig. 9.7. It has been verified by differential scanning calorimetry that the temperature for which an abrupt change occurs on the modulus curve corresponds to the glass transition temperature T_g [19]. When tests are performed at higher temperatures, the temperature of crystallisation T_c can be also determined by the abrupt increase of modulus.

T_g is so characterized by a drop of the modulus value. Below T_g the density of elastic bounds is stable and viscous movements are too slow and correlated to happen under the imposed high frequency. When these movements become faster and more independent, the density of elastic bound instantly decreases. The change of rate of the damping coefficient Q^{-1} (dissipated

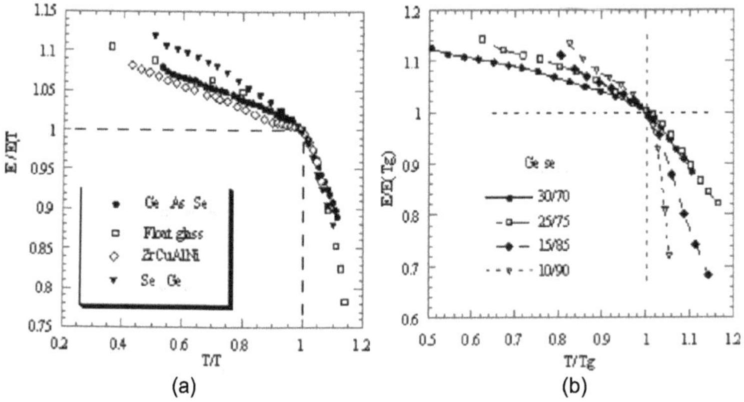

Figure 9.8 (a) Attempt of normalization of the temperature–modulus curves for different glasses. (b) Normalization of the temperature–modulus curves for different glasses of the Ge_xSe_{1-x} system.

energy/stored elastic energy ratio; see Section 9.4.1) observed in Fig. 9.7 confirms this transition [8].

The reproducibility of the "modulus versus temperature" curve shape leads to an attempt of normalization on a master curve $E/E(T_g)$ versus T/T_g [8, 20]. Figure 9.8a presents this attempt for Young's moduli varying from 17 to 89 GPa (respectively, for $Se_{70}Ge_{30}$ and for a metallic glass) and T_g varying from 280°C to 600°C (respectively, for GeAsSe and for a float glass).

We can note a slight difference between the slopes, above and below the glass transition temperature. If we assume that it does not come from experimental scattering, the study of the $Se_{1-x}Ge_x$ $(0 \leq x \leq 0.3)$ system can explain these variations. If $Se_{70}Ge_{30}$ has the structure of pure glass, the increase of the Se amount gives rise to a semipolymeric structure with Se chains. The normalization on the master curve (Fig. 9.8b) shows a decrease of the slope with the decrease of Se content. If we assume that the slope $\mathrm{d}E/\mathrm{d}T$ is proportional to the density of bounds (and to the decrease of the bounding energy), the variation of the slope can be directly attributed to the difference of elastic bounds densities between the different structures.

9.3 Elasticity of Coated Materials

9.3.1 *Single Coatings*

The mechanical optimization of substrate + structural coating(s) composites requires a good estimation of the macroscopic elasticity of each component. Moduli values relative to coatings are rather different than equivalent bulk ones for many reasons: anisotropy, cracks, porosity, texture, and boundaries' nature.

Empirical models have been earlier developed in order to use this technique for thin or thick coatings' characterization. More recently, an analytic model of composite beam vibration based on the notion of equivalent rigidity has been developed and is valid whatever the thickness of the coating [21]. Another approach based on the minimalization of the stored energy [22] gives the same results. They both only suppose that the adhesion of the coating is perfect and therefore that there is continuity of the strains at the interface.

The principle to determine the Young's modulus of a coating is to do differential measurements: the substrate resonance is firstly measured (resonance frequency F_s) and then the composite resonance measured (frequency shift ΔF). The longitudinal Young's modulus E_d of a homogeneous coating, having a density ρ_d and deposited on a substrate with a modulus E_s and a density ρ_s is given by the following equation:

$$\left(1+\frac{\Delta F}{F_s}\right)^2 = \frac{1+\varepsilon^3\left(\frac{E_d}{E_s}\right)+\frac{3\varepsilon(1+\varepsilon)^2(E_d/E_s)}{1+\varepsilon(E_d/E_s)}}{1+\varepsilon\left(\frac{\rho_d}{\rho_s}\right)} \qquad (9.6)$$

where ε is the ratio of the coating-to-substrate thickness. It has been applied to the determination from room temperature up to 1300 K of the modulus of various coatings with thickness ranging from 2 to 200 μm. Temperature stabilization is an important issue when considering the coating modulus as a function of temperature: the measured frequency shift ΔF at a given temperature must come exclusively from the coating and not from a variation of frequency of the substrate due to a slight variation of temperature. The measurement scattering is in the order of a few percent for 10–100 μm thick coatings; in the worst case with thin coatings (2 μm thick amorphous SiC and NiTi deposited on steels by dynamical ionic mixing), it reaches 15% [23].

Another interesting case is the characterization of damaged coatings illustrated with a chromium electrolytic coating presenting oriented microcracks. The apparent measured modulus E_a is in the order of 80 GPa, whereas the modulus of bulk perfect chromium is about 300 GPa [24]; a model of an elastic damaged continuous medium with a density ρ_F of parallel interacting microcracks [25, 26] allows us to establish the fitting relation

$$E_a = E\frac{1}{1 + 2\pi\rho_F e^{\pi\rho_F}} \tag{9.7}$$

In the case of coatings, if the Young's modulus can be determined with this method, it is also important to be able to evaluate other elastic parameters such as the shear modulus or Poisson's ratio. As the use of cylindrical samples in longitudinal mode (E and ν) is not adapted for coatings studies, one way to investigate is the use of torsion mode testing on plate specimens that can be easily coated. As for flexural tests, the application of Hamilton's principle minimizing potential and kinetic energies of the vibrating system and differential measurements [9] allow us to determine the shear modulus G_d of the coating:

$$\left(\frac{F_C}{F}\right)^2 \approx \frac{1 + \frac{G_d}{G_S}\varepsilon\left(3 - \frac{B}{\sqrt{3}H}\right)}{1 + \frac{\rho_d}{\rho_S}\varepsilon\frac{H^2+3B^2}{H^2+B^2}} \tag{9.8}$$

It has been applied to NiO oxides grown on Ni polycrystal characterization. This system has been choosen for several reasons: NiO is easy to elaborate regularly even on large surfaces, and this oxide is the unique Ni base oxide, ensuring a well-known density and the elaboration temperature as an influence on the oxide structure. During annealing at about 1100°C, 60 μm thick anisotropic columnar structures are developed, while oxidations at about 850°C present 40 μm thick quasi-isotropic equiaxed/columnar duplex microstructures [27].

Several samples of each microstructure have been studied to determine the shear modulus of the coating and its scattering. For 850°C oxidations, we find a Young's modulus of 106 ± 7 GPa and a shear modulus of 39 ± 3 GPa. These values are rather low due to

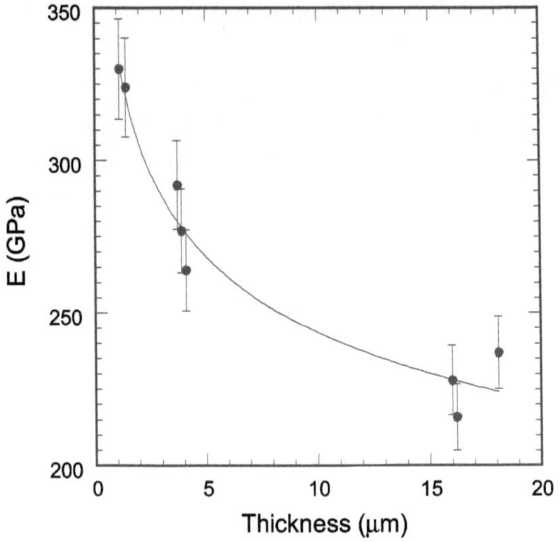

Figure 9.9 Influence of the interphase on the apparent Young's modulus of a CVD SiC coating as a function of its thickness.

porosity. In the hypothesis of isotropy, we obtain a Poisson's ratio of about 0.36. In the case of higher-temperature oxidations, these values become respectively 137 ± 2 and 38 ± 1 GPa. The scattering is lower in this case because coatings are thicker and the obtained values confirm clearly mechanical anisotropy, as expected.

9.3.2 *Homogeneous Multicoatings*

With relatively thin chemical vapor deposition (CVD) SiC coatings deposited on steels, it has be shown that the 5–10 μm thick interface introduces an artifact in the estimation of the modulus of the coating (Fig. 9.9).

The bilayer model of a composite beam has been extended analytically with the same hypothesis to a three-layer model [21] to take into account the presence of the interphase (the substrate, the interphase, and the coating characteristics are, respectively, indexed s, i, and d in the following equation). The simplified expression (first-order approximation with thin layers) linking together the resonance frequency of the whole composite (F_c) to the frequency of the substrate (F_s) is

$$\left(\frac{N_c}{N_s}\right)^2 = \frac{1 + 4\varepsilon_i \frac{E_i}{E_s} + 4\varepsilon_d \frac{E_d}{E_s}}{1 + \varepsilon_i \left(\frac{E_i}{E_s} + \frac{\rho_i}{\rho_s}\right) + \varepsilon_d \left(\frac{E_d}{E_s} + \frac{\rho_d}{\rho_s}\right)} \tag{9.9}$$

In practice, E_s is first determined with the thicker coatings applying the bilayer model and neglecting the interphase. Finally, E_i is deduced using the finer ones with the help of the three-layer model. The problem is to evaluate the density and thickness of the interphase that is not homogeneous (Fe, Si, C compounds). Different assumptions estimate an interface stiffness value between 300 and 400 GPa.

This model has been also applied to the system AM1 single-crystal superalloy coated with 45 μm anticorrosion platinum aluminide plasma spraying coating and a 125 μm zircon thermal barrier from room temperature up to 1100°C. At room temperature, the modulus scattering obtained after several runs on different specimens is, respectively, 0.2%, 2%, and 5%. Figure 9.10 shows the evolution of the modulus of each layer as a function of temperature.

If the shape of the substrate temperature–modulus curve is rather common, in the case of platinum aluminide, we can note a plateau followed by an abrupt drop at 900°C; this tendency is generally observed with plasma or high-energy deposition. Contrarily, the modulus of the columnar zircon seems to decrease faster, it certainly comes from the space created between columns due to the difference of the thermal expansions of the ceramic coating and the metal, and it introduces elastic discontinuities and the apparent modulus decreases, as described with damaged coatings.

9.3.3 *n-Layer Materials*

The last approach concerns thermochemical treatments where a gradient of mechanical properties exists from the surface to the "heart" of the material. If we suppose that the treated thickness can be considered as the stacking of n sufficiently thin homogeneous layers, the model of the composite beam can be extended to $n+1$ layers (first-order approximation). The modulus E_i $(1 \le i \le n)$ of the i layer with density ρ_i and relative thickness ε_i is given by the

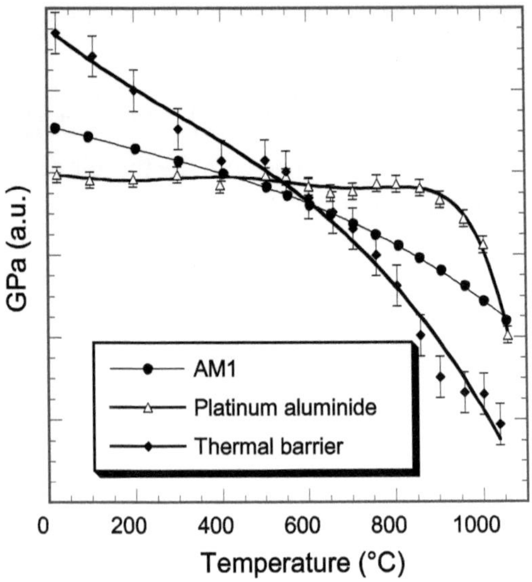

Figure 9.10 Young's modulus versus temperature curves for the three components of the system AM1 superalloy + platinum aluminide + zircon thermal barrier.

relation

$$E_{\mathrm{i}} = \frac{E_{\mathrm{s}}}{3}\left|\frac{\rho_{\mathrm{i}}}{\rho_{\mathrm{s}}} + \frac{2}{\varepsilon_{\mathrm{i}}}\left(\frac{\Delta F_{\mathrm{i}} + F_{\mathrm{s}}}{F_{\mathrm{s}}}\right)^2\right| \tag{9.10}$$

Practically, each layer must be mechanically removed and the associated frequency variation ΔF_{i} measured. To illustrate this approach, Fig. 9.11 presents the case of a nitride-hardened steel treated on 200 μm and decomposed in seven layers of about 30 μm each.

No variation of the modulus is detected as a function of depth: the average modulus of 209 GPa is equal to the modulus of the substrate (the 3% scattering introduced by successive abrasions corresponds to a maximal error of 1 μm at each removal). On the contrary microhardness measurements and internal stresses evaluation, sensitive to the presence of microprecipitates, reveal a gradient of properties. The volume of precipitates transformed during the treatment is too small to act on the global elasticity.

Another example of the elastic property gradient is given by welding; in the case of friction welding of aeronautics alloys, the

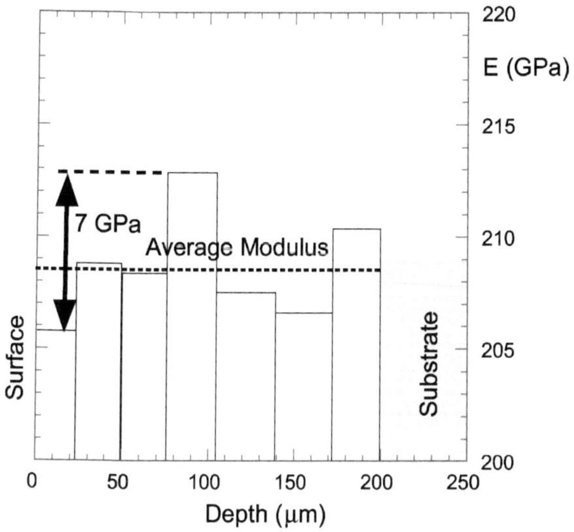

Figure 9.11 Variations of the Young's modulus of a nitride-hardened steel as a function of depth.

welding itself has a thickness of about 0.4 mm and is surrounded by two heat-affected zones (HAZs) of about 0.5 mm; with the two alloys, we obtain a five-layer system too thick to be studied. A solution is to cut on one side of the welding to obtain two three-layer systems: one reference alloy + HAZ + welding. By applying twice the analytic exact three-layer model, we can determine the modulus profile across the welding, as illustrated by Fig. 9.12.

Using the n-layer simplified model, the modulus evolution can be estimated in each layer. The scattering reaches 7% because thin layers cannot be mechanically perfectly removed due to microstructural irregularities.

9.4 Other Characterizations

9.4.1 *Anelasticity*

If for rigid materials the stress–strain linearity exists in a certain domain (up to 0.2% deformation for metals), this is not the case for flexible materials or those with strong damping. This nonlinearity is

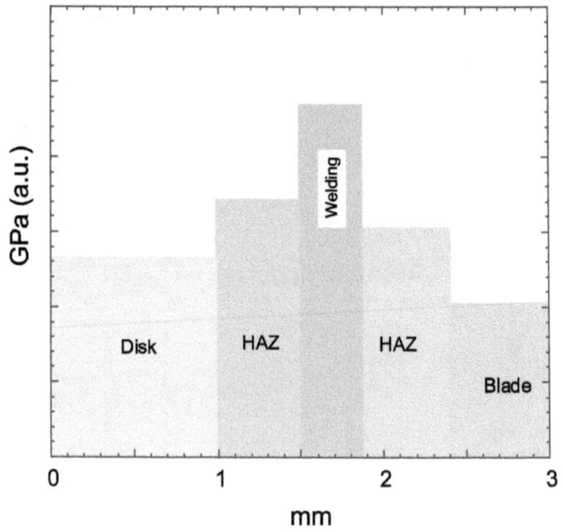

Figure 9.12 Variations of the Young's modulus across a friction welding.

also a general rule in high-temperature tests. In fact, if a macroscopic elastic limit is defined, structural defects can be mobile under a very lowamplitude load and absorb energy in a nonelastic manner; this is called *anelasticity* [28] and the damping coefficient Q^{-1} introduced in Section 9.2.3 (inverse of a quality factor) is defined by

$$Q^{-1} = \frac{1}{2\pi}\frac{\Delta W}{W} \tag{9.11}$$

where ΔW is the energy dissipated by these defects and W the stored elastic energy. This coefficient is generally quite sensitive to the loading frequency as well as to the level of loading in relation to dissipation mechanisms that are generally thermally activated

In the case of the resonant method the damping coefficient Q^{-1} (commonly called internal friction) is determined directly with a network analyzer by measuring the bandwidth ΔF of the resonance peak at −3 dB from the relation [29]

$$Q^{-1} = \frac{\Delta F}{F} \tag{9.12}$$

This coefficient, characterizing the macroscopic viscoelasticity under a low-amplitude load, is relevant to identify and analyze

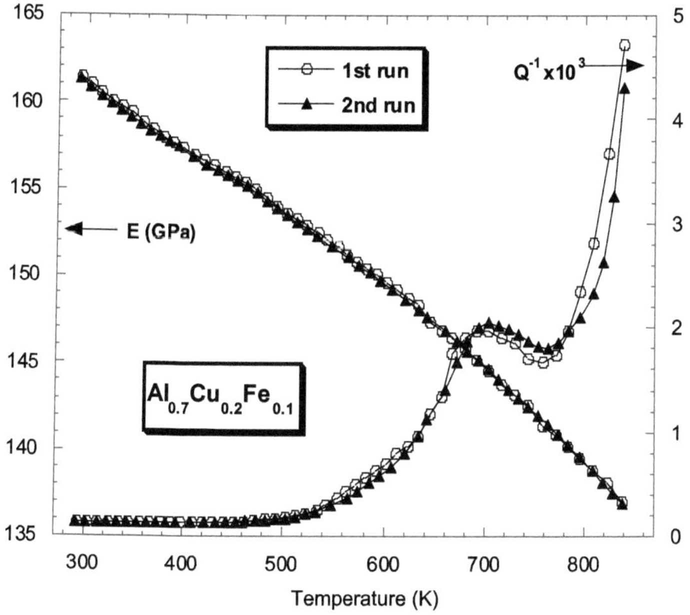

Figure 9.13 Young' modulus and damping coefficient of an ω-AlCuFe phase.

energetically structural defects at different scales (point defects, dislocations, boundaries, interfaces . . .).

We can illustrate the interest to measure this parameter with the case of a spark plasma sintering (SPS)-synthesized ω-AlCuFe phase used as a hardening phase in metal matrix composite (MMC) applications. Figure 9.13 shows the evolution of the Young's modulus and damping coefficient measured simultaneously [30] as a function of temperature

Internal friction spectra show a relaxation peak superimposed onto a high-temperature background. The relaxation peak is attributed to stress-induced reorientation of substitutional defects as observed in ordered alloys; the hightemperature background is associated with interfaces solid friction due to the specific nature of particle–particle interfaces produced by SPS elaboration. The reproducibility of the damping spectrum after annealing at 850 K indicates that the sintered microstructure is very stable.

9.4.2 *Determination of Residual Stresses*

Young's modulus and damping coefficient measurements, performed on various materials by means of the dynamic resonant method, exhibit transient effects during the first heating, while there is no obvious structural evolution. It has been more particularly observed on sintered and rolled bulk materials as well as coated materials. It can be indubitably related to the release of internal stresses introduced during elaboration. The measurement of the resonance frequency shift associated with this release and the development of a model of vibration integrating the presence of internal stresses allow the estimation of the initial level of internal stresses [22].

The case of a prestressed vibrating beam can be found in the literature [31] when a longitudinal static stress σ_s is applied: the resonance frequency varies with the amplitude of the stress. A very simple analogy can be made with the tuning of musical instrument strings.

If the static applied stress is replaced by a homogeneous longitudinal internal stress σ_R, the formalism describing the vibration is strictly the same but the sign of the strain energy is reversed because it becomes an internal energy.

If we consider a bulk sample before and after the release of internal stresses, the relation between flexural resonance frequencies as a function of the released stress is

$$\frac{F_{\sigma R}^2}{F^2} = 1 - 0.309\frac{\sigma_R L^2}{E B^2} \tag{9.13}$$

We must note that this approach takes into account the average longitudinal internal stress and can be only applied to systems where this component is preponderant. The case of isotropic stresses located at microscopic scale or of important space variations cannot be analyzed.

This effect can illustrated with a high isostatic pressure (HIP)-sintered MAX phase; sintering under high pressure ensures homogeneous macroscopic compression internal stresses. The effect of the release of these stresses is shown in Fig. 9.14. This release of 220 MPa is relatively easy and is complete at 300°C.

In the case of a 70%-rolled Co base steel having a very high elastic limit, hard forming generates anisotropy and high plane tension

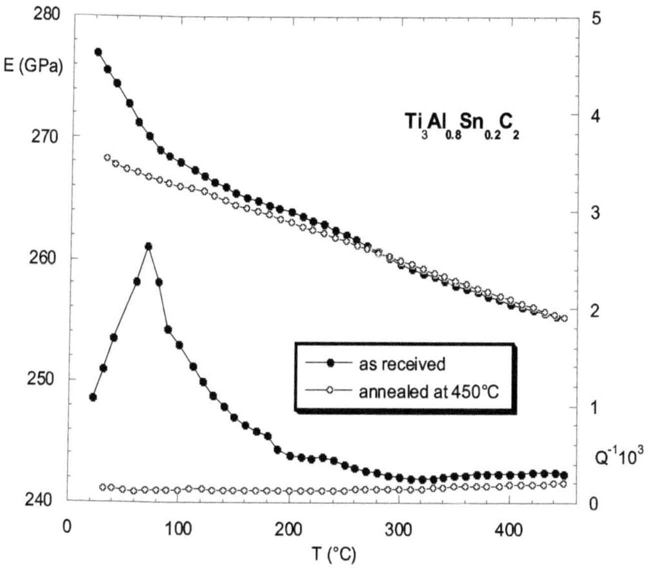

Figure 9.14 Effect of residual stresses release on Young's modulus and damping coefficient spectra.

stresses. The effect of the release is inversed (increase of modulus) because residual stresses are here in tension; the complete release is never reached even after several thermal treatments due to the confinement of the microstructure. The level of released tension stresses is about 170 MPa.

In the case of internal stresses due to coating elaboration at high temperature, it is commonly assumed that internal stresses come from the difference of the expansion coefficient between the substrate and the coating during the cooling phase (thermal stresses). It is also admitted that the level of stress is constant in the coating and that the average value in the substrate is negligible. The new relation giving the shift of resonance frequency in dynamics as a function of the released stress is

$$\frac{F_{c\sigma R}^2}{F_c^2} = 1 - 0.309\frac{\varepsilon\sigma_R L^2}{E_s B^2} \tag{9.14}$$

An illustration of this effect is given Fig. 9.15 with the system AM1 +45 μm thick anticorrosion platinum aluminide plasma spraying coating. (see Section 9.3.2). The release of tension stress in the

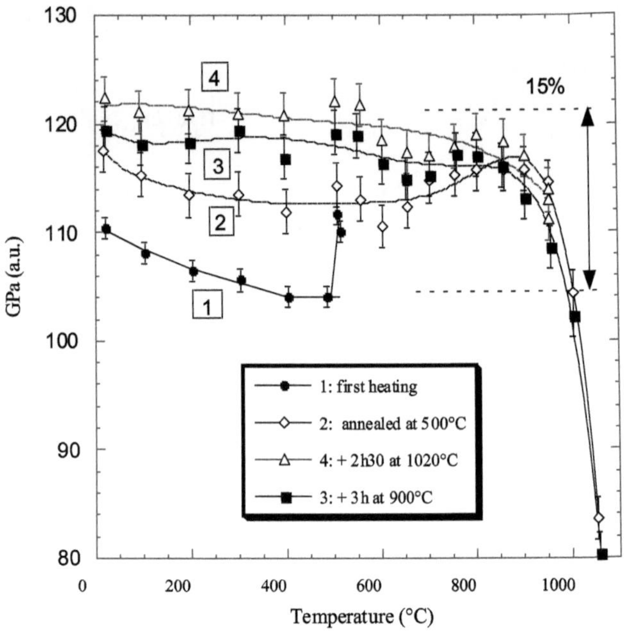

Figure 9.15 Effect of the release of thermal stresses on the modulus of a platinum aluminide coating.

coating is about 260 MPa; the frequency shift being very small, the dispersion on the value must be much higher than in the case of bulk materials. However, this value is in good agreement with X-ray diffraction analysis giving 300 MPa on other sample series.

9.5 Current Development: Elasticity of thin Films

All the experimental and analytic developments made for the characterization of coatings' elasticity are limited to the study of 1–2 μm thick coatings. As this determination comes from differential measurements, the coating/substrate thickness ratio cannot be too low to obtain accurate results. To study submicronic films, we must so take substrates in the order of 0.1–0.2 μm thick (instead of the

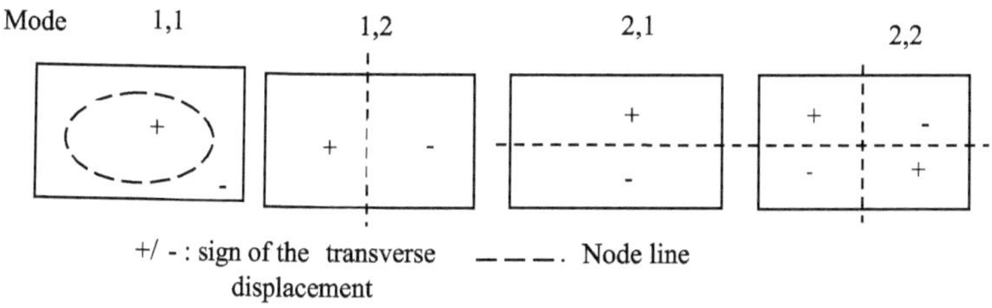

Figure 9.16 First harmonics shape for plate flexural vibrations.

1–2 μm thickness range employed for flexural or torsional tests) to get a reasonable ratio. But with these dimensions, we do not obtain flexural beam vibrations (one-dimensional) but plate vibrations (two-dimensional) and the analytic formalism (Eqs. 9.1 and 9.6) is not valid.

For these reasons, we develop currently a new testing head to study the vibration of plates in a view to characterize the elasticity of thin films. As the equation of vibration of a plate has only an analytic solution when it is simply supported on its four edges, this kind of support has been firstly tested but it presents instabilities in dynamics (transverse vibrations due to the too low inertia of the sample). A second possibility is to clamp the plate on one edge but this kind of support is strictly banned by ASTM standards [4] (imperfect boundary conditions). The suitable solution is to excite the plate in free flexural mode. For that the testing head with crossed wires used for torsional vibrations (see Fig. 9.1) can be employed with any change, whereas the position of the excitation detection electrode must be shift. As the fundamental resonance frequency (1,1 mode) implies geometry-dependent boundary conditions, the solution is to use 2,2 mode (see Fig. 9.16) for which orthogonal crossed wires ensure optimal boundary conditions.

The last work is to obtain the analytic expression of the thin bulk or coated plate flexural vibrations using the same approach as previously exposed: application of Hamilton's principle minimizing potential and kinetic energies described by the Lagrangien of the system [9, 22].

References

1. Spinner, S., and Valore, R.C. (1958). Comparison of theoretical and empirical relations between the shear modulus and torsional resonance frequencies for bars of rectangular cross section, *J. Res. ASTM: A Phys. Chem.*, **60**, 459–464.
2. Spinner, S., Reichard, T.W., and Tefft, W.E. (1960). A comparison of experimental and theoretical relations between Young's modulus and the flexural and longitudinal resonance frequencies of uniform bars, *J. Res. ASTM: A Phys. Chem.*, **64A**, 147–155.
3. Spinner, S., and Tefft, W.E. (1961). Torsional resonance vibrations of uniform bars of square section, *J. Res. ASTM*, **65A**, 167–171.
4. ASTM (2001). Standard test method for Young's modulus, shear modulus and Poisson's ratio by impulse excitation of vibration, in *Annual Book of ASTM Standards*, **03.01**, 1099–1112.
5. Nuovo, M. (1961). Mesures dynamiques de constantes élastiques et anélastiques, *La Ricerca Sci.*, **1**, 212–243.
6. Glandus, J.C. (1981). Rupture fragile et resistance aux chocs thermiques de ceramiques a usage mecanique (Thèse, Université de Limoges, France).
7. Mazot, P., de Fouquet, J., Woirgard, J., and Pautrot, J.P. (1992). Mesure des constantes d'élasticité et du frottement intérieur des solides par méthode de résonance entre 20 et 1200°C, *J. Phys. III*, **2**, 751–763.
8. Gadaud, P., and Pautrot, S. (2003). Characterization of the elasticity and anelasticity of bulk glasses by dynamical subresonant and resonant techniques, *J. Non-Cryst. Solids*, **316**, 145–152.
9. Gadaud, P., Milhet, X., and Pautrot, S. (2009). Bulk and coated materials shear modulus determination by means of torsional resonant method, *Mater. Sci. Eng. A*, **521–522**, 303–306.
10. Timoshenko, S., and Goodier, J.N. (1951). *Theory of Elasticity* (McGraw-Hill).
11. Davies, R.M. (1938). The frequency of longitudinal and torsional vibration of bars, *Philos. Mag.*, **25**, 364–386.
12. Pickett, G. (1945). Equations for computing elastic constants from flexural and torsional resonant frequencies, *Am. Soc. Test. Mater. Proc.*, **45** 846–865.
13. Covarel, G. (2001). Etude par spectrométrie mécanique isotherme d'alliages CuAlNi et relation avec la microstructure (Thèse, Université de Poitiers, France).

14. Bozorth, R.M. (1993). *Ferromagnetism* (IEEE Press, Piscataway).
15. Rouxel, T. (2007). Elastic properties and short-to medium-range order in glasses, *J. Am. Ceram. Soc.*, **90**, 3019–3039.
16. Gueguen, Y., Gadaud, P., Rouxel, T., et al. (2011). High-temperature elasticity and viscosity of Ge_xSe_{1-x} glasses in the transition range, *Phys. Rev. B*, **84**, published online.
17. Gadaud, P. (2008). *Structural Components-Mechanical Tests and Behaviour Laws* (Wiley Editions).
18. Pautrot, S., Lagattu, F., and Gadaud, P. (2003). Influence des renforts sur l'évolution en température du module d'Young de différents composites à matrice polymère, *Ann. Chim.*, **28**, 43–52.
19. Le Bourhis, E., Gadaud, P., Guin, J.P., Tournerie, N., Zhang, X.H., Lucas, J., and Rouxel, T. (2001). Temperature dependence of the mechanical behaviour of a GeAsSe glass, *Scripta Mater.*, **45**, 317–323.
20. Gadaud, P., and Pautrot, S. (2004). Application of the dynamical flexural resonance technique to industrial materials characterization, *Mater. Sci. Eng.*, **A370**, 422–426.
21. Mazot, P., and Pautrot, S. (1998). Détermination du module d'Young de dépôts par flexion dynamique: application aux systèmes bicouche et tricouche, *Ann. Chim. Sci. Mater.*, **23**, 821–827.
22. Gadaud, P., and Pautrot, S. (2012). Determination of residual stresses by means of dynamic resonant method, *Solid State Phenom.*, **184**, 461–466.
23. Peraud, S., Pautrot, S., Villechaise, P., and Mendez, J. (1997). Determination of the Young's modulus by a resonant technique applied to two dynamically ion mixed thin films, *Thin Solid Films*, **292**, 55–60.
24. Pautrot, S., and Mazot, P. (1993). Mesures par méthode dynamique du module d'Young du dépôt de chrome sur un acier allié, *Rev. Métall.*, 1665–1667.
25. Halm, D., and Dragon, A. (1998). An anisotropic model of damage and frictional sliding for brittle materials, *Eur. J. Mech. A: Solid*, **17**, 439–460.
26. Kachanov, M. (1992). Effective elastic properties of cracked solids, *ASME Appl. Mech. Rev.*, **45**, 304–335.
27. Milhet, X., Cormier, J., Renault, P.O., Coupeau, C., and Colin, J. (2005).Stress heterogeneity of thermally grown polycrystalline nickel oxide layers, *Mater. Sci. Eng.*, **A395**, 22–26.
28. Zener, C. (1948). *Elasticité et Anélasticité des Métaux* (Dunod).
29. Nowick, A.S., and Berry, B.S. (1972). *Anelastic Relaxation in Solids* (Academic Press, New York).

30. Laplanche, G., Gadaud, P., Bonneville, J., Joulain, A., Gauthier, V., and Dubois, S. (2011). SPS sintering synthesis and mechanical spectroscopy of ω-AlCuFe phase, *J. Mater. Sci.*, published online.
31. Géradin, M., and Rixen, D. (1994). *Mechanical Vibrations* (Wiley-Masson).

Index